AF737800

Environmental Remediation Technologies, Regulations and Safety

Gulf Coast Ecosystem Restoration Strategy and Long-Term Recovery Plan

Environmental Remediation Technologies, Regulations and Safety

Additional books in this series can be found on Nova's website under the Series tab.

Additional E-books in this series can be found on Nova's website under the E-book tab.

Environmental Science, Engineering and Technology

Additional books in this series can be found on Nova's website under the Series tab.

Additional E-books in this series can be found on Nova's website under the E-book tab.

ENVIRONMENTAL REMEDIATION TECHNOLOGIES, REGULATIONS AND SAFETY

GULF COAST ECOSYSTEM RESTORATION STRATEGY AND LONG-TERM RECOVERY PLAN

BRADLEY F. BROWN

AND

CODY B. DAVIS

EDITORS

Nova Science Publishers, Inc.

New York

Copyright © 2012 by Nova Science Publishers, Inc.

All rights reserved. No part of this book may be reproduced, stored in a retrieval system or transmitted in any form or by any means: electronic, electrostatic, magnetic, tape, mechanical photocopying, recording or otherwise without the written permission of the Publisher.

For permission to use material from this book please contact us:
Telephone 631-231-7269; Fax 631-231-8175
Web Site: http://www.novapublishers.com

NOTICE TO THE READER

The Publisher has taken reasonable care in the preparation of this book, but makes no expressed or implied warranty of any kind and assumes no responsibility for any errors or omissions. No liability is assumed for incidental or consequential damages in connection with or arising out of information contained in this book. The Publisher shall not be liable for any special, consequential, or exemplary damages resulting, in whole or in part, from the readers' use of, or reliance upon, this material. Any parts of this book based on government reports are so indicated and copyright is claimed for those parts to the extent applicable to compilations of such works.

Independent verification should be sought for any data, advice or recommendations contained in this book. In addition, no responsibility is assumed by the publisher for any injury and/or damage to persons or property arising from any methods, products, instructions, ideas or otherwise contained in this publication.

This publication is designed to provide accurate and authoritative information with regard to the subject matter covered herein. It is sold with the clear understanding that the Publisher is not engaged in rendering legal or any other professional services. If legal or any other expert assistance is required, the services of a competent person should be sought. FROM A DECLARATION OF PARTICIPANTS JOINTLY ADOPTED BY A COMMITTEE OF THE AMERICAN BAR ASSOCIATION AND A COMMITTEE OF PUBLISHERS.

Additional color graphics may be available in the e-book version of this book.

LIBRARY OF CONGRESS CATALOGING-IN-PUBLICATION DATA

Gulf coast ecosystem restoration strategy and long-term recovery plan / editors, Bradley F. Brown and Cody B. Davis.
p. cm.
Includes index.
ISBN 978-1-61942-781-5 (hardcover)
1. Restoration ecology--Gulf Coast (U.S.) 2. Ecosystem management--Gulf Coast (U.S.) 3. Coastal ecology--Gulf Coast (U.S.) 4. Coastal zone management--Gulf Coast (U.S.) 5. Oil spills--Management--Gulf Coast (U.S.) 6. Gulf Coast (U.S.)--Environmental conditions. 7. Gulf Coast (U.S.)--Economic conditions. 8. Gulf States--Environmental conditions. 9. Gulf States--Economic conditions. I. Brown, Bradley F. II. Davis, Cody B. III. Gulf Coast Ecosystem Restoration Task Force (U.S.)
QH104.5.G84G83 2011
639.90976--dc23

2011053198

Published by Nova Science Publishers, Inc. † New York

CONTENTS

PREFACE

The Deepwater Horizon oil spill was one of the worst man-made environmental disasters our country has ever experienced. The oil spill dramatically affected the lives, jobs, and futures of millions of Gulf Coast residents. The Gulf of Mexico is a natural resource of vital importance which provides immeasurable benefits and services to citizens throughout the United States. The Gulf is also critical to nationwide commerce. This book examines the Gulf Coast Ecosystem Restoration Strategy and Long Term Recovery Plan with a focus on facilitating long-term vitality through restoring and conserving habitat, restoring water quality, replenishing and protecting living coastal and marine resources, and on making the Gulf of Mexico a healthier and more resilient place for the communities that depend on its wealth of resources.

In: Gulf Coast Ecosystem Restoration Strategy ...
Editors: Bradley F. Brown and Cody B. Davis
ISBN: 978-1-61942-781-5
© 2012 Nova Science Publishers, Inc.

Chapter 1

GULF OF MEXICO REGIONAL ECOSYSTEM RESTORATION STRATEGY*

Gulf Coast Ecosystem Restoration Task Force

I. EXECUTIVE SUMMARY

The Gulf of Mexico is among the nation's most valuable and important ecosystems. This ecosystem, consisting of offshore waters and coastal habitats of Alabama, Florida, Louisiana, Mississippi and Texas, is home to ecologically, culturally, commercially and recreationally important species of fish and wildlife. The Gulf of Mexico is an ecosystem composed of both coastal and marine environments and resources that contribute to its function, health and productivity as a single, interconnected system. The Gulf ecosystem and its natural resources are important to the U.S. economy, producing 30 percent of the nation's gross domestic product in 2009.[1] The region provides more than 90 percent of the nation's offshore oil and natural gas production,[2] 33 percent of the nation's seafood,[3] 13 of the top 20 ports by tonnage in the United States in 2009,[4] and significant recreation and tourism benefits. The five U.S. Gulf Coast states, if considered an individual country, would rank seventh in global gross domestic product.[5] The Gulf Coast region's economy is highly intertwined with its natural resource base, including oil and gas deposits, commercial and recreational fisheries, waterfowl, migratory birds and other wetland-dependent life, coastal beaches, and waterways for ports, waterborne commerce and tourism.[6]

The four overarching goals for Gulf Coast restoration are to:

- Restore and conserve habitat
- Restore water quality
- Replenish and protect living coastal and marine resources
- Enhance community resilience

* This is an edited, reformatted and augmented version of the Gulf Coast Ecosystem Restoration Task Force, dated December 2011.

Although the Gulf Coast has significant natural resource and economic value, its long-term future is not secure. Gulf Coast states have experienced coastal land loss due to the alteration of natural hydrology and other human activities, as well as from events such as tropical storms and hurricanes. The building of levees on the Mississippi River and its tributaries since the 18th century has contributed to depriving once-thriving wetlands and barrier islands of the freshwater, sediments and nutrients they need to survive.[7] Sediment is essential to the Gulf ecosystem, where coastal Louisiana was formed over the course of 7,000 years by deltaic processes, including intermittent flooding of the Mississippi River, which delivered sediment to different coastal regions. In the Gulf, communities, commerce and ports have long relied on the extensive network of human-made flood damage risk reduction and navigation structures for their existence. Accordingly, river management priorities historically have centered on navigation and flood risk reduction. While clearly successful in meeting those two goals, those priorities created unintended consequences to the surrounding environment by accelerating wetland and barrier island erosion and restricting the flow of vital sediments that had sustained the Gulf ecosystem over time. Oil and gas industry canals, pipelines and other infrastructure crisscross the landscape to accommodate exploration, development and commercial activity related to these enterprises; they also contribute to wetland loss. Furthermore, studies indicate that geologic land subsidence exacerbates sea-level rise impacts.[8] Over the last seven decades, approximately 1,883 square miles of wetlands in Louisiana alone have been lost—a land mass roughly the size of Delaware.[9]

The Gulf of Mexico watershed covers 56 percent of the continental United States (see "Gulf of Mexico Watershed Extent in the United States").[10] The Mississippi River Basin alone drains an estimated 40 percent of the continental United States.[11] Excess nutrients (nitrogen and phosphorus) and pollutants flowing into the Gulf from upstream basins degrade water quality in the Gulf of Mexico. As a result, the Gulf's estuaries have become increasingly degraded for both human use and aquatic life.

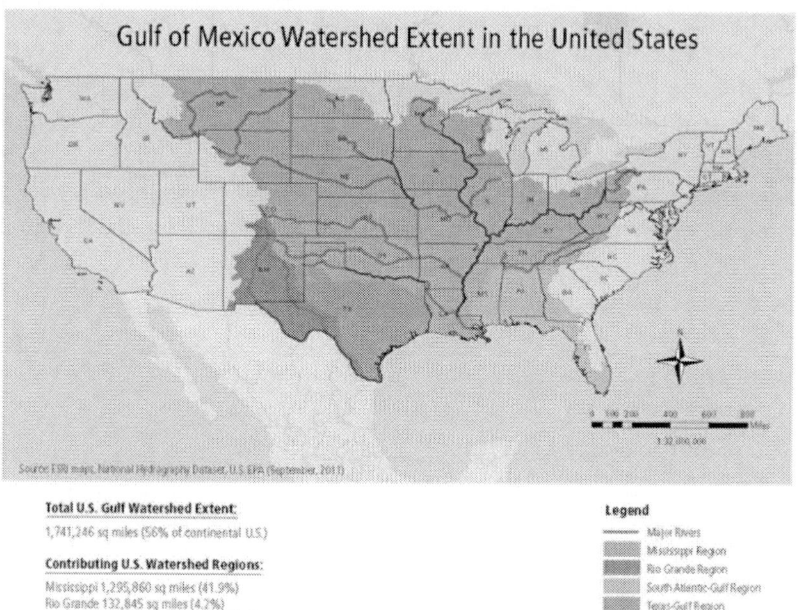

Recognizing the importance of the Gulf of Mexico and its ecosystems, and in response to the *Deepwater Horizon* oil spill and the recommendations proposed by Secretary of the Navy Ray Mabus in *America's Gulf Coast: A Long Term Recovery Plan After the Deepwater Horizon Oil Spill* (September 2010), President Barack Obama established the Gulf Coast Ecosystem Restoration Task Force on October 5, 2010.[12] The purpose of the task force is to coordinate the long-term conservation and restoration of America's Gulf Coast. The task force consists of senior officials from seven federal cabinet agencies, the Executive Office of the President, and representatives from the five Gulf Coast states of Alabama, Florida, Louisiana, Mississippi and Texas. As part of its mandate, the task force is charged with developing a Gulf of Mexico Regional Ecosystem Restoration Strategy to drive action and guide the long-term collaboration that will be necessary to effectively address and reverse widespread environmental degradation and to ensure a healthy environment and economic future for the Gulf. This Strategy considers the important observations and recommendations proposed in the National Commission on the BP *Deepwater Horizon* oil spill and offshore drilling report to the President: *Deep Water, The Gulf Oil Disaster and the Future of Offshore Drilling*, published January 2011 (also called the Oil Spill Commission Report).[13]

The Strategy: Overarching Goals and Framework for Restoration

Many of the Gulf's challenges are complex and long-standing, and addressing the problems will require sustained and consistent effort over time. The task force's Strategy was developed through a review of existing plans and efforts and significant public outreach during the last year. The Strategy builds on ongoing work and priorities of each of the Gulf Coast states, local communities, federal partners, academics and nongovernmental organizations (NGOs). The restoration framework consists of four overarching goals to guide the collective actions at the local, state and federal levels that are necessary to reverse the ongoing decline and restore the Gulf Coast's ecosystem. The task force's goals for the Gulf Coast restoration effort are the following:

- Restore and conserve habitat
- Restore water quality
- Replenish and protect living coastal and marine resources
- Enhance community resilience

The Strategy includes enhanced community resilience as a goal in order to highlight the connectedness of ecosystem restoration with the well-being and sustainability of coastal communities. This goal also emphasizes the planning and technical assistance needs of vulnerable coastal communities as they assess risks, plan for and rebound from natural disasters, and implement sustainable development approaches. As progress is made to improve water quality and protect and enhance the Gulf of Mexico's habitats and resources, communities benefit from improved ecosystem services, such as better storm protection and healthier fisheries.

A key function of the task force is to help advance state, federal and tribal activities that will achieve the overarching restoration goals. The task force intends to help member agencies address process and other barriers, facilitate program implementation and alignment, and better leverage scientific expertise and fiscal resources in support of Gulf Coast restoration.

In addition to encouraging better collaboration among state and federal agencies and tribal entities, partnerships play an integral role in restoring ecosystem functions throughout the Gulf region. The task force believes it is critically important to foster an inclusive dialogue and expand on public-private partnerships in order to achieve ecosystem restoration in the Gulf of Mexico.

Major Actions

Within each of the four goals, the task force has identified specific actions that require the most immediate attention, and that will build the initial foundation for an effective regional partnership framework. These are articulated in the following table:

Major Actions	
Goal: Restore and Conserve Habitat	• Prioritize ecosystem restoration in the Gulf of Mexico by ensuring that social, environmental and economic outcomes are fully considered in all river management decisions, and by placing it on equal footing with other priorities such as navigation and flood damage risk reduction. • Improve current sediment management practices to maximize to the extent practicable and ecologically acceptable the quantity and effective use of sediments by taking a "strategic use" approach to sediment management. • Restore and preserve more natural river processes of sediment and freshwater distribution. • Expand the network of state, federal and private conservation areas to ensure healthy landscapes that support the environment and culture of the region and the diverse services provided by the Gulf of Mexico ecosystem. • Restore and conserve coastal and near-shore habitats, with a focus on marshes, mangroves, seagrasses, barrier islands, natural beaches and dunes, and coastal forests and prairies.
Goal: Restore Water Quality	• Decrease and manage excess nutrient levels in the Gulf through the development and implementation of state nutrient reduction frameworks. • Focus restoration actions in priority watersheds to address excess nutrients in coastal waters and reduce hypoxic conditions. • Reduce pollutants and pathogens from stormwater flows and other sources. • Improve the quality and quantity of freshwater flow into priority estuaries to protect their health and resiliency. • Coordinate and expand existing water quality monitoring efforts supporting adaptive management of programs and projects designed to improve water quality. • Collaborate with Mexico to assess and reduce emissions from oceangoing vessels in the Gulf that degrade water quality.
Goal: Replenish and Protect Living Coastal and Marine Resources	• Restore depleted populations of living coastal and marine resources. • Conserve and protect offshore environments. • Restore and protect oyster and coral reefs, and other coastal environments. • Coordinate and expand existing Gulf monitoring efforts to track sentinel species and sites. • Minimize, and eliminate where possible, invasive species that impact the Gulf of Mexico.
Goal: Enhance Community Resilience	• Develop and implement comprehensive, scientifically based, and stakeholder-informed coastal improvement programs. • Provide analytical support tools to enhance community planning, risk assessment and smart growth implementation. • Enhance environmental education and outreach.

The task force recognizes the value of intergovernmental collaboration, the need for dedicated funding for large landscape-level restoration efforts, and the importance of a strong scientific foundation for restoration. Accordingly, the task force reiterates recommendations made by Secretary Mabus that call for Congress to:

- Formalize the long-term intergovernmental partnership among the Gulf Coast states and the federal government by establishing a successor coordinating body to the task force.
- Dedicate a significant portion of the eventual Clean Water Act civil penalties resulting from the *Deepwater Horizon* oil spill for Gulf recovery, in addition to current funding for Gulf programs.

The priorities and recommended actions outlined in the Task Force Strategy should be used to guide the application of resources directed to Gulf ecosystem restoration by Congress. Since Congress has not yet fully addressed these recommendations, this Strategy is built on existing authorities and resources. It is a significant step representing enhanced collaboration and a recognition of the shared responsibility among federal and state governments to restore the Gulf Coast ecosystem. In this time of severe fiscal constraints, task force member agencies and states are committed to establishing shared priorities and working together to achieve them.

The Strategy details a restoration framework and series of actions that task force member agencies can take to support each of the restoration goals. It also lays out a series of next steps that will better align agency programs and leverage scientific and fiscal resources. Given the inherent complexities of undertaking ecosystem restoration on this scale, a key focus of implementation will be to ensure that ecosystem restoration efforts have a robust scientific foundation and use an effective adaptive management framework, including necessary monitoring, modeling, research and decision-support tools. Integrating adaptive management into restoration efforts will help ensure that progress can be made, even in the absence of scientific certainty. The implementation component of the Strategy will set forth specific on-the-ground actions and milestones by which the intergovernmental partners can achieve and measure progress of the restoration of America's Gulf Coast.

II. Introduction

Importance and Value of the Gulf of Mexico

Spanning nearly 600,000 square miles across five U.S. states (Alabama, Florida, Louisiana, Mississippi and Texas), six Mexican states, and Cuba, the Gulf of Mexico constitutes a diverse and vibrant ecosystem, which is a vital environmental, economic and cultural asset for the United States. Despite many significant environmental and human-made stressors, the Gulf is able to support a host of commercial and recreational uses and provide the backdrop for the unique cultures and heritage of this region. However, unless bold and broad-scale measures are taken soon, the health and future of the Gulf will remain in jeopardy.

The Gulf of Mexico constitutes a diverse and vibrant ecosystem which is a vital environmental economic and cultural asset for the United States.

The Gulf is endowed with a variety of coastal and marine habitats, including wetlands, barrier islands, beaches, and coral and oyster reefs. These habitats are integral to the economies and cultural fabric of the Gulf and the nation, providing a range of ecosystem services including fisheries, wildlife-related activities, food production, energy production, infrastructure protection and recreational opportunities. Healthy Gulf Coast habitats also contribute to the resilience of Gulf Coast communities, providing a line of defense for coastal communities and their associated infrastructure against powerful storms. The Gulf's wetlands provide a natural flood attenuation function, which may reduce the impacts of flooding associated with storms. Healthy wetlands could also reduce potential future impacts associated with climate change.

Gulf habitats are rich havens of biodiversity. The Gulf Coast's wetlands, beaches, coastal woodlands and waterbird nesting islands are major nurseries for breeding birds, and provide foraging and stopover sites for millions of migrating birds that converge from several of the most important migratory flyways. Coastal marshes and near-shore habitats provide essential nursery habitat for ecologically, commercially and recreationally important species of fish and invertebrates. Offshore, the Gulf supports biologically diverse marine habitats and species, including deepwater corals, sponges, fish stocks and other unique communities. The Gulf region is also home to coastal, marine and freshwater species listed as threatened or endangered and several species of protected marine mammals.

The Gulf Coast is a powerful economic engine for the nation and home to a wide range of industries, including more than 90 percent of offshore oil and gas production,[14] one-third of the U.S. seafood harvest,[15] and a vast network of commercially important shipping lanes and ports. Tourism and recreational activities, such as fishing, boating, beachcombing and bird watching, support more than 800,000 jobs[16] across the region, making a significant economic input to Gulf communities and the nation. All of these industries depend on a healthy and resilient Gulf.

Problems Affecting the Gulf

The *Deepwater Horizon* oil spill was a reminder of the delicate balance among the environment, the economy and public health in the region. However, the oil spill was only the most recent in a long line of negative environmental impacts (e.g., hypoxic zone, and effects from navigation canals and energy pipelines) that have plagued the Gulf states for decades. Following the oil spill, Secretary of the Navy Ray Mabus led an assessment of the most pressing challenges facing the Gulf of Mexico ecosystem. The report, *America's Gulf Coast: A Long Term Recovery Plan After the Deepwater Horizon Spill* (September 2010), described them as follows:[17]

- *Loss of wetland habitats, including coastal marshes, forested wetlands, barrier islands, and coastal shorelines that form the Mississippi River Delta and Chenier Plains.* While an issue in every Gulf state, the loss of coastal habitat has been most dramatically illustrated in Louisiana and highlights the need to maintain freshwater and sediment flows to the Gulf of Mexico. Since the 1930s, the coast of Louisiana has lost over 2,000 square miles (25-35 square miles per year) of wetlands. Causes of this loss include a combination of erosion, storm damage, land subsidence, alterations to natural freshwater and sediment flow from the Mississippi River, dredging of canals for oil and gas exploration and pipeline installation activity, and the construction of navigation and flood control structures along the Mississippi River. Climate change (including the impacts of inundation and sea-level rise) threatens to accelerate the loss of these habitats.

- *Erosion of barrier islands and shorelines throughout the Gulf Coast.* From Florida to Texas, continued erosion of the coastal barrier island system undermines storm protection for coastal communities, threatens the beaches that support the local tourism economy, and affects numerous species that rely on these barrier islands for habitat (e.g., Kemp's Ridley and loggerhead sea turtles, numerous shorebirds and the Alabama beach mouse).
- *Loss and degradation of coastal estuarine habitat.* The estuaries and coastal systems of the Gulf Coast—such as Mobile Bay, Apalachicola Bay, Galveston Bay, Tampa Bay, Florida Bay, the Mississippi Sound, Barataria Bay and others—provide the nursery habitat for most of the fishery resources in the Gulf, and support a nationally important oyster industry. These estuaries are impacted by a variety of stressors, including pollution, coastal development, energy development, erosion, hydrological alteration, changes in freshwater inflow, structural marsh management and overfishing. Many of these bay systems have been recognized as estuaries of national significance by the National Estuary Program of the Environmental Protection Agency (EPA) under the Clean Water Act or as National Estuarine Research Reserves under the Coastal Zone Management Act, and are the focus of existing place-based protection and restoration efforts.
- *Imperiled fisheries.* Several of the major commercially and recreationally important finfish species are currently experiencing pressures from overfishing or have been overfished. In some cases, these conditions have persisted for many years. Additionally, contaminants such as methyl-mercury in fishes, and red tide organisms and human pathogens in shellfish, reduce fishery values and endanger human health. Under the Magnuson-Stevens Fishery Conservation and Management Act, federally managed populations of National Oceanic and Atmospheric Administration (NOAA) trust species such as red snapper, grouper and mackerel are being rebuilt through the efforts of the Gulf of Mexico Fishery Management Council and state fishery management agencies. The impact of the *Deepwater Horizon* oil spill on the rebuilding efforts for these species is as yet unclear.
- *Hypoxia (low oxygen) in the Gulf of Mexico.* Hypoxia occurs where the concentration of dissolved oxygen in the water column decreases to a level that reduces the quality of habitat, resulting in death or migration away from the hypoxic zone. The northern Gulf of Mexico adjacent to the Mississippi River is the site of the largest hypoxic zone in the United States and the second largest hypoxic zone worldwide. This Gulf of Mexico "Dead Zone" is caused by input of excess nutrient pollution to the Gulf most of which comes from upstream through Mississippi River drainage. A federal-

state Hypoxia Task Force has been working to address factors leading to low-oxygen conditions, and EPA and the Department of Agriculture (USDA) have jointly worked to develop strategies to reduce nutrient runoff. NOAA has been working on developing models to support better understanding of biological-systems transport, including hypoxia predictions.

- *Climate change.* Our changing climate is already altering, perhaps irreversibly, the physical, chemical and biological characteristics of our oceans, coasts and adjacent watersheds. Increasing air and water temperatures, changing precipitation patterns, rising sea level, and ocean acidification will increasingly confound efforts to restore or sustain system states. Federal and state natural resource managers need the information and tools to develop strategies for mitigating and adapting to a dynamic environment and ongoing habitat reorganization and restructuring.

Other environmental impacts are evident in localized areas of the Gulf. For example:

- Bays and estuaries throughout the Gulf of Mexico, such as Galveston Bay in Texas and Apalachicola Bay in Florida, have been degraded by reduced freshwater input from the upstream construction of dams and the heavy use of river water for municipal, industrial and agricultural purposes.
- Channels and canals for navigation and oil and gas activity in Texas and Louisiana have allowed a greater inflow of salt water from the Gulf of Mexico into the estuaries. In some cases, this has caused saltwater intrusion into freshwater marshes and forested wetlands, stressing these habitats and converting them to open water.
- In Mississippi, past dredging of the Pascagoula River and its economically vital navigation channel has interrupted the natural transport of sediments required for the nourishment of the barrier islands. Recent changes in sediment management practices to include downdrift placement in the littoral zone have minimized current sediment loss; however, this is not sufficient to allow for barrier island recovery.
- In Alabama, the five rivers emptying into Mobile Bay carry excess sediment, which results in high turbidity, or clouding of the water. The clouded water deprives submerged aquatic vegetation of the light it needs to grow. This unique vegetation provides important habitat for marine species and waterfowl.
- In Florida, bay systems near Pensacola, Tampa, Naples and Fort Myers suffer from near-coastal water quality issues, such as periodic harmful algal blooms (HABs) affecting marine life, human health and tourism.
- Populations of key plant species that are culturally significant to a number of federally recognized tribes within the Gulf States have been lost over time. For example, since European settlement of the Gulf, longleaf pines have declined from occupying 90 percent to occupying approximately 3 percent of forests within the Gulf region.[18]

The degradation of coastal and marine environments and the services they provide illustrates the often unintended consequences associated with managing the effects of well-intended and necessary projects such as navigation and flood control in such a complex system. Future impacts associated with storms, land subsidence, sea-level rise and river system management will only amplify the region's vulnerability. Scientific research, models, forecasts and visualization tools will inform steps to curb the impacts of multiple stressors on Gulf ecosystems to reduce the harm to the marine and terrestrial environment, national commerce, maritime industry, energy security, fisheries and cultures of the Gulf Coast. Steps taken to achieve one goal often have a direct impact on others. Solutions to these complex challenges also transcend state, national and international boundaries. This creates not just an opportunity but also a requirement to increase collaboration and communication to more fully understand the Gulf of Mexico and the work that is necessary for its overall restoration.

Role of the Gulf Coast Ecosystem Restoration Task Force

On October 5, 2010, President Barack Obama issued Executive Order 13554 (see Appendix A), establishing the task force. The task force was created on the recommendation of Secretary of the Navy Ray Mabus in his report on the *Deepwater Horizon* oil spill. Ecosystem restoration, as defined by the Executive Order, means:

> ... all activities, projects, methods, and procedures appropriate to enhance the health and resilience of the Gulf Coast ecosystem, as measured in terms of the physical, biological, or chemical properties of the ecosystem, or the services it provides, and to strengthen its ability to support the diverse economies, communities, and cultures of the region. It includes activity that initiates or accelerates the recovery of an ecosystem with respect to its health, integrity, and sustainability. It also includes protecting and conserving ecosystems so they can continue to reduce impacts from tropical storms and other disasters, support robust economies, and assist in mitigating and adapting to the impacts of climate change.

The task force is directed to address the persistent and significant decline of the Gulf ecosystem by working with state and federal agencies, tribes, communities, stakeholders and the public throughout the Gulf Coast to develop an ecosystem restoration Strategy.

Strategy Development

Building on prior federal and state efforts, the task force developed a shared vision for the Gulf Coast, which is to achieve a resilient, healthy Gulf of Mexico ecosystem that supports the diverse economies, communities and cultures of the region. The vision requires that the federal government and states align their activities effectively and collaborate to set realistic and measurable goals for restoration and protection of wetlands, barrier islands, and other coastal and marine features. While this Strategy focuses on the restoration of the Gulf ecosystem, it supports and builds on the first line of defense—protection—provided by other essential efforts for maintaining existing areas of healthy habitat across the Gulf Coast.

During the course of their work, task force members witnessed first-hand local and state restoration activities and visited with many groups and associations integral to Gulf restoration. The mission of the task force was not to develop another "new" plan. Rather, the task force set out to build on existing work, learn from those who are actively involved in ecosystem restoration, and craft an agenda that would provide unified and strategic direction for restoration activities across the Gulf. Building on the region's existing work, the task force aims to jump-start action to bring about a healthier and more vibrant Gulf ecosystem.

Public and Stakeholder Engagement

In developing this Strategy, the task force met at least once in each of the five Gulf states, beginning in Pensacola, Florida, on November 8, 2010, and ending in Biloxi, Mississippi, on August 30, 2011. Each task force meeting included public listening sessions to gather valuable individual input from those most connected to the Gulf. Initially, the task force designed sessions to generate discussion on specific ecosystem restoration focus areas, and to gather individual ideas and opinions from particular stakeholder groups, including local governments, business and industry, academics and NGOs. Later task force meetings focused on specific aspects of the Strategy and included panel presentations by experts in water quality, community resilience, habitat conservation, public engagement and science. In addition to the listening sessions that took place during public meetings, the task force held multiple listening sessions throughout the Gulf with partner organizations such as the National Estuary Programs (NEPs), local government leadership and academic institutions. Approximately 2,000 attendees contributed to Strategy development through task force listening sessions.

The task force also looked to two of EPA's federal advisory committees, the Local Government Advisory Committee (LGAC) and the National Environmental Justice Advisory Council (NEJAC), for input into Strategy development. Both groups were responsive in mobilizing work groups to generate views and recommendations to the EPA Administrator, Lisa P. Jackson, who shared this information with members of the task force. In May 2011, based on public input received, Administrator Jackson announced plans to establish a new EPA federal advisory committee, a 25-member Gulf of Mexico Citizen Advisory Committee, which will become an additional vehicle for citizen engagement and support as the Strategy recommendations are implemented.

Coordination on Health and Economic Concerns

From the first task force meetings, public audiences have included many directly affected by the *Deepwater Horizon* oil spill. Public comments from these meetings included oil-spill-related concerns, such as economic claims problems, health-related issues, and seafood safety questions, among others. While the Strategy focuses on ecosystem restoration, the task force has shared these concerns relating to the spill with the relevant departments and agencies.

During the course of task force meetings, federal, state and local government officials provided the task force with updates on seafood safety issues and oil spill recovery activities. In addition, representatives from the Department of Health and Human Services and the Department of Commerce were routinely invited to participate in task force public listening sessions to meet individually with and answer questions from members of the public on these matters.

Consultation and Coordination with Tribes

The task force conducted consultation and coordination with affected federally recognized tribes in Gulf Coast states during the development of the Strategy. EPA facilitated the task force consultation following Executive Order 13175 on consultation and coordination with tribes and EPA's Tribal Consultation Policy, and was also informed by the consultations supporting the oil spill response and Natural Resource Damage Assessment (NRDA) process. The tribal consultation and coordination included face-to-face meetings, webinars and conference calls. Tribes had access to background documents and could provide comments via EPA's Tribal Portal (www.epa.gov/tribal). All comments were summarized and provided via the Tribal Portal. In response to discussions with tribal representatives from the federally recognized tribes, the task force also sought input from tribes with ancestral lands in the Gulf.

Science to Inform Gulf Ecosystem Restoration Strategy

Recognizing that successful ecosystem restoration must be based on sound science, the task force established teams to examine specific topics that would support the Strategy's restoration goals. These teams sought to identify both the best science available and the critical gaps in our knowledge of the Gulf ecosystem. Specifically, the Task Force Science Coordination Team examined current conditions within the Gulf ecosystem and science needs to help support restoration and conservation. Additionally, task force member agencies and states have contributed their extensive expertise and perspectives working on other ecosystem restoration efforts (e.g., the Comprehensive Everglades Restoration Plan, or CERP) to address the long-standing challenges facing the Gulf, including how best to design and implement restoration projects and assess their success using a robust scientific foundation and an adaptive management framework.

Future Efforts for the Task Force

Leveraging Partnerships

Throughout the five Gulf states, partnerships among communities, NGOs, private industry, foundations, landowners and government agencies have played an integral role in the success of prior and ongoing conservation and restoration efforts. The task force believes it is critically important to foster an inclusive dialogue and expand public/private partnerships in order to successfully implement the goals of this Strategy. Particularly in times of challenging budgets, partners are an essential component of achieving ecosystem restoration. The task force intends to engage with communities, private industry, foundations, landowners and NGOs in order to leverage each partner's investments and identify activities for maximum benefit to the ecosystem.

To optimize efficiency, the task force plans to continue to assess efforts and capabilities—its own and others'—to ensure that it is adding value to restoration activities in the Gulf region. The task force operates in coordination with multiple intergovernmental bodies, including the Gulf of Mexico Alliance (GOMA), the Mississippi River/Gulf of Mexico Watershed Nutrient Task Force (Hypoxia Task Force), the NRDA Trustee Council for the *Deepwater Horizon* oil spill, and the National Ocean Council. Each of these entities provides expertise and capacity to address a unique group of problems and issues in the Gulf of Mexico. The task force seeks to improve cooperation and coordination among various federal, state, tribal and local entities, reduce duplication of efforts, and help align and move forward sustainable resource management strategies, restoration plans and resiliency projects.

- *The Gulf of Mexico Alliance.* One group that shares important goals for the future of the Gulf is GOMA, which was formed in 2004 with the goal of significantly increasing regional collaboration to enhance the ecological and economic health of the Gulf of Mexico. This regional partnership includes the governors of the states of Alabama, Florida, Louisiana, Mississippi, and Texas, as well as federal partners, including EPA, NOAA, USDA, the U.S. Department of the Interior (DOI), the U.S. Department of Defense, the National Aeronautics and Space Administration (NASA), the U.S. Department of Transportation (DOT), the Council on Environmental Quality, the U.S. Department of State, the U.S. Department of Health and Human Services, the U.S. Department of Energy, and the National Science Foundation. To complement state and federal leadership and provide additional expertise and resources, GOMA has strong established partnerships with academia, Gulf research institutions, industry and NGOs. The six GOMA priority issue working teams focus on: (1) water quality for healthy beaches and seafood, (2) habitat conservation and restoration, (3) ecosystems integration and assessment, (4) reducing nutrient impacts to coastal ecosystems, (5) coastal community resiliency, and (6) environmental education.[19] GOMA achieves its mission through strong and strategically leveraged Gulf-wide interstate and cross-federal-agency research integration, data sharing, conservation and restoration policy analysis, regional sediment management, and ecosystem-based habitat protection and restoration efforts.
- *The Hypoxia Task Force.* The Hypoxia Task Force includes members from 17 state and federal agencies. EPA currently co-chairs the task force with Mississippi. The mission of the Hypoxia Task Force is to understand the causes and effects of hypoxia in the Gulf; coordinate activities to reduce the size, severity and duration of the hypoxic zone; and ameliorate its effects. It provides a forum for state water quality and agriculture agencies to partner on local, state and regional efforts to mitigate nutrient loading, encouraging a holistic approach that takes into account upstream sources and downstream impacts.

  In its most recent Gulf Hypoxia Action Plan (2008), the Hypoxia Task Force emphasized its commitment to work with states to develop nutrient reduction strategies and increase accountability.

  In 2012, Iowa will assume the state co-chair position through 2014.
- *Natural Resource Damage Assessment (NRDA) Trustee Council.* The work of the task force is distinct from, but related to, the ongoing work of the Trustee Council, which is conducting a NRDA for the *Deepwater Horizon* oil spill. The Trustee Council focuses on assessing the *Deepwater Horizon* oil-spill-related natural resource injuries. The task force has a broader charge, extending to other long-standing challenges facing Gulf Coast ecosystems beyond the *Deepwater Horizon* oil spill. These challenges include significant coastal land and wetland loss, degraded water quality, depletion of marine resources, coastal erosion and the adverse effects

of climate change. The task force is directed to devise an agenda for the long-term restoration and conservation of the diverse ecosystems of the Gulf Coast that will ensure its long-term environmental, economic and health benefits. Additionally, the task force was directed to foster collaboration among governments, the public and diverse stakeholders to build on existing plans and activities designed to address Gulf restoration. Collaboration and coordination will be a key ongoing responsibility of the task force in its work to facilitate implementation of state and federal ecosystem restoration programs.

- *The National Ocean Council.* The National Ocean Council and the task force, both established by Presidential Executive Order, share important elements for the future of the Gulf Coast. The National Ocean Council is charged with implementing the National Ocean Policy and addressing broad, national enhanced stewardship of our oceans, coasts and Great Lakes, including economic, environmental, social and national security issues. The task force is charged with addressing ecosystem restoration needs in the Gulf of Mexico region and supports implementation of an important piece of the overarching National Ocean Policy.

The National Ocean Council and the task force are both engaged in promoting regional-scale protection and restoration, addressing water quality impacts and other large-scale threats, ecosystem-based management, and providing coordination and support among federal and state agencies. The National Ocean Council will reinforce and support the task force on regional and geographic initiatives and activities through its national strategies. Furthermore, the Gulf of Mexico Regional Ecosystem Restoration Strategy will inform the National Ocean Council's national strategic action plans. The overlap of federal agency members of both bodies will ensure that they will share best practices, information and advancements in science and management of coastal ecosystems.

The Council and the task force will work together to better integrate and coordinate planning, decision-making, and regulatory enforcement and ensure the integration of best practices, information, discoveries, and advancements in science and management of coastal ecosystems. These efforts will promote and sustain a culture of shared stewardship, both across federal agencies and among federal, tribal, state and local jurisdictions. Both efforts share a commitment to openness, transparency and a bottom-up, regionally driven approach to planning and action, with opportunities for tribal, state and local governments; the private sector; academic and nonprofit stakeholders; and the public to engage and help define the way forward. These connections will help ensure a coordinated effort in addressing challenges and opportunities for the Gulf region, and jointly incentivize the effective, efficient and economical restoration of the Gulf Coast's environmental, economic and cultural value.

Promoting Science-Based Decision-Making

The federal government and states must be able to leverage limited resources in a manner that yields maximum benefits to the ecosystem, consistent with this Strategy. Ultimately, successful implementation of the Strategy must be based on the best available science to inform management decisions. Priorities should be implemented according to the principle of adaptive management, allowing for restoration efforts to move ahead while addressing the need to decrease uncertainty and expand and incorporate knowledge of the ecosystem conditions. Adaptive management will help determine the efficacy of restoration actions through a focused effort of monitoring, modeling and research to support effective management and decision-making. This process will help ensure that state and federal investments can be altered if they are not achieving the desired results, ultimately improving the overall effectiveness of restoration and protection efforts. Leveraging the successes of existing state and federal programs and initiatives, addressing critical gaps, and building on the capacity that currently exists within the Gulf of Mexico states are central to supporting adaptive management. Specific needs for successful adaptive management are discussed in Section IV, "Science-Based Adaptive Management."

Resolving Policy and Process Obstacles Impeding Progress

The task force evaluated barriers that have hindered implementation and success of past restoration efforts. The majority of plans developed in the region in the last 20 years document the same stressors outlined in this Strategy. A primary focus of the Strategy will be to address the policy and procedural barriers complicating progress on protection, conservation and restoration efforts.

Some of these barriers were identified during task force meetings with stakeholders and the public. Budget constraints of federal, state and local governments are a barrier to ecosystem restoration, delaying or restricting restoration efforts. Additionally, inadequate coordination within and among federal and state agencies impedes movement of projects to construction. Water resource policies also inhibit some ecosystem restoration efforts, such as those designed to use sediment for greater ecosystem benefits. Furthermore, inconsistent or unclear priority-setting undermines cooperation and support for projects. And, limited research and science challenge many aspects of planning, priority-setting and project design.

A greater emphasis on cooperative research, monitoring and data sharing is needed to improve project planning and design and to support adaptive management over time. Alternative financing mechanisms should be explored and implemented to foster progress on restoration projects. Future solutions may involve statutory changes to address obstacles that cannot be overcome through administrative remedies. In the near term, existing resources in regulation and policy need to be fully exercised in order to ensure greater alignment of federal and state actions.

The Strategy Going Forward

This Strategy embraces and builds upon existing efforts at the state, federal and tribal levels. It is a significant step, representing enhanced collaboration and a recognition of the shared responsibility among state, federal and tribal governments to restore the Gulf Coast ecosystem. In this time of severe fiscal constraint across all levels of government, task force member agencies are committed to finding common ground, establishing priorities, and working together to achieve them. This may involve reassessing budgets and agency activities to collaboratively align resources to the highest priority Gulf Coast restoration work. This Strategy will be an important tool in the alignment of the agencies' resources needed to carry out implementation.

Gulf Coast States

The Gulf Coast is rich in natural resources that drive economic activity in the area. The beauty and bounty of the coast draws millions of visitors from across the nation and around the world each year, creating a multibillion-dollar tourist industry that contributes both jobs and revenues. Each of the five states boasts a strong fishing industry, and some of the busiest ports in the nation are found on the Gulf Coast. The area is the heart of the U.S. oil and gas industry. Important military bases, space centers and research facilities are also found in the region, increasing the prosperity of individual states, tribes and the nation as a whole.

Although each state has features that make it unique and distinct, the Gulf's interrelated ecosystem connects its people and communities. Together the states face many threats, both natural and human-made. For all of the Gulf states, the ecosystem and its restoration are vital to the region's continued economic well-being.

A brief description of each state's coastal area is provided below, along with a summary of its major environmental and economic assets. Additionally, Appendix B contains a more detailed description of each state's natural resources and the environmental services and economies they support. Appendix B also describes each state's priority actions for the four overarching goals included in this Strategy.

Alabama

Coast[20]	Tidal Shoreline[21]
53 miles	607 miles
State Population[22]	**Gulf Coast Population[23]**
4,779,736	764,613

Alabama's coastal area is located along the northern Gulf of Mexico. The state has a profuse diversity of natural habitats, including rich sediments, seagrass beds, barrier islands, wetlands, plant bogs, bottomland hardwood forests, wet pine savannas and pine and oak forests. The wealth of these habitats makes the Alabama coast unique and draws residents and visitors alike. In terms of biodiversity, Alabama ranks first among the states east of the Mississippi River and fifth among all states in the nation. The Mobile Bay Watershed is the sixth largest river basin in the United States and the fourth largest in terms of stream flow.[24] The Alabama coastal area contributes to the economy of the state and the entire nation in several important ways. These include:

- *Commercial and recreational fishing.* In 2009, commercial fishing trips landed 29.7 million pounds of seafood worth over $40.5 million. Commercial fishing contributes

$391 million and over 8,750 jobs to the state's economy annually.[25] Alabama's recreational fishing increases these figures to a yearly economic impact of $865 million and more than 13,680 jobs being created.[26] Seventy-five percent of America's commercial fish catch and 80 to 90 percent of the recreational fish catch are dependent on estuaries such as Alabama's Mobile Bay, Mississippi Sound, Weeks Bay and Perdido Bay.[27]

- *Tourism.* In 2009, more than 7.1 million visitors enjoyed the state's beaches, scenic vistas, access to the Gulf of Mexico, and bays and waterways, spending more than $3.2 billion in Mobile and Baldwin Counties. The counties rank first (Baldwin at 25 percent) and third (Mobile at 9.7 percent) in total travel-related employment for the state, with 56,294 employees.[28]
- *Shipping and maritime services.* The Port of Mobile is a global deepwater gateway for the state. The port ranked 12th in the nation in total tonnage shipped in 2009.[29] Combined with the city of Mobile's industrial center, the port represents a multi-billion-dollar impact on regional, state and world economies. The port contributes 66,617 jobs and $7.92 billion annually to Alabama's economy.
- *Energy production.* Alabama is rich in energy resources and has considerable natural gas resources. The area's shipbuilders are supporting offshore oil and gas exploration by building offshore supply and rig-tending vessels and repairing rigs.

MIGRATORY BIRDS ON DAUPHIN ISLAND

Dauphin Island is one of the top birding spots in the Southeast. An incredible 347 species have been reported on the island. The island is the first landfall for many migratory birds making the 600-mile flight across the Gulf of Mexico from the Yucatan Peninsula. Strong rains and winds during these flights can trigger spectacular "fallouts" in which large flocks of many bird species descend on the island to seek shelter.

Source: Dauphin Island Bird Sanctuaries. <http://www.coastalbirding.org>

Florida

Coast[30]	Tidal Shoreline[31]
770 miles	5,095 miles
State Population[32]	**Gulf Coast Population[33]**
18,801,310	7,771,053

Florida's barrier islands, estuaries, coral reefs, beaches, seagrass meadows, coastal wetlands and mangrove forests are world-renowned natural resources and attractions. Along the Gulf Coast alone, Florida has more than 7.4 million acres of tidally submerged lands.[34] The state's sandy beaches are consistently ranked among the best in the nation, and millions of residents and visitors alike come to Florida's Gulf coast each year to fish, dive, swim, and view wildlife. These natural resources also represent Florida's economic engine for the future and support a range of activities, including:

- *Tourism.* In 2010 tourism was responsible for welcoming more than 82.3 million visitors to Florida, who spent more than $62.7 billion, generating 22 percent of the state's sales tax revenue and employing nearly a million Floridians.[35] Each year, almost $15 billion and 141,373 jobs come directly from fish and wildlife in the state, and nearly another $17 billion and another 202,743 jobs are the indirect benefit of boating activities in state waters.[36]
- *Wildlife viewing.* Wildlife viewing is a significant pastime in Florida, accounting for $5.2 billion and 51,367 jobs in 2006. Also in 2006, almost 1.6 million participated in wildlife viewing in Florida, the majority of whom came to view coastal and marine life.[37]
- *Shipping and cruise services.* Florida ships agricultural and industrial products and transports cruise ship customers through its many Gulf Coast ports—Pensacola, Panama City, Port St. Joe, St. Petersburg, Tampa, Port Manatee and Key West. These Gulf Coast ports account for about $10.6 billion per year in economic activity, supporting 125,000 direct and indirect jobs.[38]
- *Military installations.* Training and testing operations along Florida's Gulf Coast provide significant conservation lands and economic and security benefits to the state and nation. Statewide, defense-related spending was $64.8 billion in 2010 and accounted for 686,181 jobs.[39]
- *Commercial and recreational fishing.* In 2009, Florida's commercial fishing industry ranked second among all states for annual in-state sales ($13 billion) and third for jobs (64,744 jobs).[40] Also in 2009, the state ranked tenth in total landings at $116 million annually and produced 10 percent of the Gulf's oyster catch (by weight), a $6.9 million annual dockside value.[41] Florida also leads all states in economic return for its marine recreational fisheries. Recreational saltwater fishing generates over $5 billion and more than 50,000 jobs each year.[42] In 2008–2009, more than one million people bought marine recreational fishing licenses, a third from out of state. More than 3,400 for-hire fishing licenses were purchased, generating more than $1 million and giving Florida one of the largest charter fleets in the world.[43]

The Beauty and Bounty of Florida's Coral Reefs

Extensive coral reefs off Florida's Gulf Coast support an incredible diversity of fish and other marine life. The beauty and biodiversity of the state's coral reefs attract tourists, fishermen and divers from around the world. The state, particularly the Florida Keys, is recognized as a diving capital of the world.

Louisiana

Coast[44]	Tidal Shoreline[45]
397 miles	7,721 miles
State Population[46]	**Gulf Coast Population[47]**
4,533,372	3,548,090

Louisiana has the largest expanse of coastal wetlands in the lower 48 states and is home to the largest delta in North America.[48] Louisiana's coast provides a home to nearly half of the state's population. The state's ecosystems include wetlands, shorelines, and adjacent forested and scrub/shrub areas. The Mississippi River system has shaped Louisiana's uniquely formed coastal zone, which is dominated by wetlands and filled with sensitive resources. These wetlands truly are "America's wetlands." They are important to Louisiana citizens, as well as the nation, who depend on them for a variety of important activities, including:

- *Shipping and maritime services.* The Mississippi River system significantly contributed to the growth of this nation. Today, more than 30 states depend on coastal Louisiana's navigation channels for imports and exports. In 2009, Louisiana's ports accounted for 20 percent of all waterborne commerce (by tonnage) in the nation.[49] Five of the 15 largest ports in the United States are in Louisiana.[50] The state's top three ports handle an estimated 55 to 70 percent of all U.S.-exported corn, soy and wheat. The state's ports also handle other commodities such as steel, coffee, rubber and chemicals.[51]
- *Energy production.* Louisiana's offshore waters are home to the largest reserves of oil and gas in the nation, producing over 90 percent of the nation's Outer Continental Shelf oil and 70 percent of the country's Outer Continental Shelf natural gas.[52] Louisiana is the top producer of oil and the second-highest producer of natural gas in the nation (including Outer Continental Shelf oil and gas).[53] The U.S. Treasury benefits by up to $8 to $10 billion annually from offshore energy revenues attributable to waters offshore of Louisiana.[54] This constitutes one of the largest contributions to the treasury.
- *Commercial and recreational fishing.* Louisiana's coast includes one of the nation's largest and most productive estuaries. Nearly 26 percent (by weight) of the commercial fish and shellfish in the continental United States are harvested in Louisiana's waters.[55] Louisiana is the nation's top producer of shrimp, oysters, crawfish and blue crabs. Coastal Louisiana is also home to some of the top recreational fishing waters (both fresh and salt water) in the country. Ninety-seven percent (by weight) of the commercial fish and shellfish harvested in the Gulf of Mexico are species that depend on coastal wetlands for reproduction, hatching growth or

development. Louisiana alone has approximately 40 percent of the coastal wetlands in the continental United States.[56]

- *Tourism.* Nearly 25 million domestic and international tourists visited Louisiana in 2010, adding an estimated $9.3 billion in spending to the nation's economy.[57] Nearly half of this spending was in the New Orleans area.

DUCKS FIND PRIME WINTERING GROUNDS ON LOUISIANA'S COAST

More than five million ducks and other water birds (20 percent of the continental population) winter in Louisiana each year. During the peak spring migration season, nearly 25 million neotropical migrant passerines and shorebirds arrive in coastal Louisiana each day.

Source: Louisiana Department of Wildlife and Fisheries. 2011. Waterfowl population estimates in Louisiana's coastal zone below U.S. Highway 90 and on Catahoula Lake. <http://www.wlf.louisiana.gov/sites/default/files/pdf/waterfowl_survey/33575-January%202011%20Survey/waterjan2011.pdf>

Mississippi

Coast[58]	**Tidal Shoreline[59]**
69 miles60	359 miles
State Population[61]	**Gulf Coast Population[62]**
2,967,297	628,502

Mississippi's coastal waters encompass the mainland coast, barrier islands, bays and lagoons, as well as hundreds of rivers, creeks and estuaries. The state's largest estuary is the Mississippi Sound, which encompasses 550 square miles.[63] The state's estuaries and streams are home to a remarkable variety of plants and animals in what is recognized as one of the most biologically diverse regions in North America. The area ranks in the top 10 for native species of reptiles, amphibians, butterflies and mammals.[64] The culture and economy of the Mississippi Gulf Coast are heavily influenced by the industries that rely on the abundance of natural resources for their viability. These include:

- *Commercial fishing and seafood processing.* These industries are a natural extension of life in this coastal area. Hundreds of fishing boats make their home in Mississippi ports, and the seafood they catch and process generates thousands of business opportunities.

- *Tourism.* Mississippi attracts tourists through its recreational fishing opportunities, robust gaming industry, white sand beaches, championship golf courses and other recreational and cultural venues. Tourism in the three counties that border the Gulf Coast accounts for about $1.6 billion in visitor expenditures, 32 percent of state travel and tourism tax revenues, and 24,000 direct jobs.[65]
- *Energy production.* Offshore oil and gas exploration and development further boost the area's economy. Providing experienced offshore workers, necessary logistical support and industrial capacity to generate that support is an important source of fuel for the coast's economic engine.
- *Shipping and maritime services.* The Mississippi State Port at Gulfport is an economic force for jobs and business activity. The port generates more than 2,000 jobs for Mississippi residents, with that number expected to increase substantially with improvements and expansion as the city undergoes a comprehensive recovery from Hurricane Katrina. Pascagoula is the home of the largest military shipbuilder in the United States, and the largest private employer in the state, providing approximately 11,000 jobs for residents of the northern Gulf region.[66]
- *Space center.* NASA's Stennis Space Center is home to more than 30 federal, state, academic and private organizations and numerous technology-based companies. The center's staff includes approximately 2,000 oceanographers, scientists and support staff with a unique capability to study the Gulf Coast from space and in the field.

RESERVE PROVIDES CRITICAL HABITATS

The 18,000-acre Grand Bay National Estuarine Research Reserve is one of the largest, relatively undisturbed estuarine marsh/pine savanna habitats remaining along the northern Gulf of Mexico. Straddling the Mississippi-Alabama state line, the reserve provides critical habitats for many kinds of migratory birds as well as important commercial and recreational species of fish.

Texas

Coast[67]	Tidal Shoreline[68]
367 miles	3,359 miles
State Population[69]	Gulf Coast Population[70]
25,145,561	8,287,623

The Texas coastal zone includes a complex system of barrier islands and peninsulas, with Padre Island being the longest undeveloped barrier island in the world. The Texas coast contains 12 distinct eco-regions and is richly endowed with natural resources, including sand dunes, vast wetlands and aquatic habitats. The Texas coast has *tremendous* biodiversity. More than 457 species of fish and 343 species of invertebrates are found in the state's estuarine and marine waters. The coast is home to blue crabs, oysters, pelicans, plovers, shrimp, and the endangered whooping crane and Kemp's Ridley sea turtle. The ecologically rich coastal system supports many industries, including:

- *Shipping and maritime services.* Roughly one-third (423 miles) of the Gulf Intracoastal Waterway (GIWW) is located in Texas. The GIWW is the nation's third busiest waterway. Thirteen percent of the nation's freight transported by ship (74 million tons of cargo) traveled the Texas GIWW in 2006, with an estimated value of $25 billion.[71,72] The waterway handles 90 percent of all gasoline shipped to the Lower Rio Grande Valley, and facilities on the Texas Gulf Coast produce two-thirds of the nation's petrochemicals.[73] The Texas coast is also home to four of the top 10 ports in the country (based on total cargo tonnage),[74] which generate more than $9 billion in federal tax revenues annually.
- *Energy production.* The area is home to the nation's largest concentration of oil refineries[75] as well as a sizable chemical industry, which is ranked first in the nation in size and production.[76]
- *Commercial and recreational fishing.* The state's commercial fishing fleets bring in more than $150 million of fish and shellfish annually.[77] Eighty-two percent of shrimp in the United States come from the Gulf states, with Texas supplying 89.7 million pounds per year.The annual oyster harvest is approximately 5.7 million pounds of meat worth over $19 million. The recreational fishing industry is another important part of the Texas coastal economy, with saltwater sport fishing generating almost $2 billion annually.[78] The number of annual saltwater fishing permits increased more than 7 percent between 2006 and 2010.[79]
- *Tourism.* Tourists visiting the Texas coast spend more than $7.5 billion annually for beach recreation, bird watching, fishing and eco-tourism.[80] The coast accounts for more than one-quarter of total travel expenditures in Texas, making the coast the second most popular tourist destination in the state.[81]

THE RICH RESOURCES OF THE TEXAS COAST

Nearly two-thirds of the state's Gulf shoreline is protected in parks, wildlife refuges, and natural areas off-limits to development, Texas currently operates 93 state parks and natural areas, 50 wildlife management areas, and eight fish hatcheries. These comprise over 1.4 million acres that are managed in the public trust for recreation and conservation.

III. GOALS

This restoration Strategy builds upon existing research, planning and program efforts throughout the Gulf that have generated wide interest and participation by Gulf-based citizens, businesses, scientists, industries and governments. Beginning with its first meeting on November 8, 2010, the task force solicited individual input from the general public and engaged key stakeholder groups throughout the region. Additionally, the task force reviewed numerous publications that address Gulf of Mexico restoration and engaged the broad expertise of the member agencies. This input led the task force to identify the following four goals as priorities for Gulf of Mexico ecosystem restoration:

- *Restore and conserve habitat.* The Gulf Coast has endured extensive damage to key coastal habitats such as wetlands, coastal prairies and forests, estuaries, seagrass beds, natural beaches and dunes, and barrier islands. Within this goal, a major focus is to work with Gulf Coast stakeholders to expedite implementation and improve the effectiveness of state and federal programs related to landscape-scale resource management, habitat conservation and restoration strategies.
- *Restore water quality.* The Gulf of Mexico experiences numerous water quality problems, including excess nutrients, altered sediment inputs, pathogens, and mercury and other pollutants. One of the most prevalent signs of such problems in the Gulf of Mexico is hypoxia—low oxygen levels in the water—which can result from excess nutrients in the water and other factors. Within this goal, a major focus is to reduce the amount of nutrients flowing into the Gulf and to undertake other measures to enhance water quality.
- *Replenish and protect living coastal and marine resources.* Living coastal and marine resources are showing visible signs of distress, such as depleted species populations and degraded habitats. Within this goal, a major focus is to promote sustainable resource management that focuses on actions to conserve and restore viable populations of living coastal and marine resources and their coastal and offshore environments.
- *Enhance community resilience.* Gulf Coast communities face a number of pressing challenges, such as storm risk, sea-level rise, land loss, depletion of natural resources, and compromised water quality. Within this goal, a major focus is to integrate the creation of resilient communities with ecosystem restoration through the development of comprehensive coastal planning programs.

This Strategy articulates a series of actions across these four goals. These goals—just like the Gulf of Mexico ecosystem, with its connected processes, functions and elements—are intertwined. The goals, recommendations and actions outlined below are designed to organize the Strategy in a manner that can be easily understood, while at the same time acknowledging the holistic and integrated nature of the Gulf ecosystem.

Restore and Conserve Habitat

The Gulf Coast faces significant habitat restoration and conservation challenges. These challenges include habitat loss from increased development, resource management, alterations to hydrology and sediment transport, land subsidence, erosion, sea-level rise, hurricanes and tropical storms. Unless the rapid rate of coastal land and habitat loss in the region is halted and eventually reversed, the ecosystem and the services it provides could collapse. This collapse would yield negative consequences for the marine and terrestrial environment, national commerce, the maritime industry, energy security, fisheries, and the rich cultural legacy of the Gulf Coast region. In order to restore the Gulf ecosystem, federal, tribal, state and local governments should work to address, moderate and reverse the factors that drive the degradation of Gulf Coast habitats in a scientifically defensible manner. Working with private partners, landowners and NGOs, governments should strive for a sustainable and resilient ecosystem where wetlands, estuaries and barrier shorelines are managed to achieve and maintain a productive balance between ecological, economic and social functions.

A science-based freshwater and sediment management regime coupled with large-scale restoration has significant potential to address some of the leading causes of coastal habitat losses. It should be implemented in a way that ensures the long-term survival and sustainability of the unique coastal cultures and the resource-based livelihoods of the region. The implications of no action would further threaten the future of a number of coastal communities and would lead to a continued decline in ecosystem services. Key ecosystem functions can be protected and enhanced by reconnecting rivers with their deltaic plains and managing or reestablishing freshwater and sediment inflows while maintaining the effectiveness of navigation and flood damage risk reduction. Healthy ecosystems provide functions that improve coastal resilience. A comprehensive watershed-based approach to the management of river systems is required to ensure that current and future ecosystem needs are met. This comprehensive approach would help ensure that these areas are managed for maximum benefit to both the natural and human systems of the region.

Giving ecosystem restoration equal footing with navigation and flood damage risk reduction is an important element of this Strategy and should be applied to river management activities across multiple agencies. Such an approach would ensure that economic, environmental and social needs are appropriately considered and weighed as part of a decision-making processes. Water resource investment decisions at all levels of government have relied too heavily on the economic outcomes and have not appropriately balanced environmental and social outcomes, regardless of the type of action (navigation, flood damage reduction, ecosystem restoration, water treatment, etc.). Given the limited resources at all levels of government, it is important that a broader array of outcomes be considered with the goal of maximizing public benefits at both site-specific and watershed scales. These new decision-making processes should include consideration of disproportionate impacts and outcomes often faced by economically and socially vulnerable populations from natural hazards and environmental degradation. Through a consideration of a broader array of outcomes, these processes will appropriately advance important ecosystem restoration actions in the Gulf, without adversely impacting actions needed to address flood damage reduction, navigation or other such water resources based needs.

The following major actions are recommended to restore and conserve Gulf Coast habitat.

Major Actions

Prioritize ecosystem restoration in the Gulf of Mexico by ensuring that social, environmental and economic outcomes are fully considered in all river management decisions, and by placing it on equal footing with other priorities such as navigation and flood damage risk reduction

The management of river systems has for too long been focused on addressing issues such as navigation and flood damage risk reduction without commensurate attention to the environmental and social needs of the region. In order to address the resulting coastal sediment distribution issues, stem habitat loss, improve coastal community resilience, and restore water quality to achieve comprehensive restoration, a more modern approach to the management of river systems is needed to ensure that solutions to current and emerging problems can be developed holistically. As a first step and to protect and restore the Gulf of Mexico, stem habitat loss and restore water quality, coordinated action at all levels of government to redefine river management priorities in consideration of a full array of benefits to the public is necessary. To this end, the task force should:

- Review all federal, tribal and state authorities, capabilities, plans and activities underway relevant to river management and evaluate the contribution of each to the long-term goal of ecosystem restoration of the Gulf of Mexico.
- Following consultation with member agencies, develop recommendations for needed authorities, funding priorities and policy changes—including actions that may be undertaken within existing authorities.
- Provide input from a regional perspective to agencies developing project implementation guidance that reflects changes to water resources policy; work with agencies to incorporate this guidance, which will be designed to promote more effective, efficient and holistic future use of water resources.

Improve current sediment management practices to maximize to the extent practicable and ecologically acceptable the quantity and effective use of sediments by taking a "strategic use" approach to sediment management

Sediment, delivered by the Gulf river systems, built much of the Gulf Coast and continues to be essential to the health of the Gulf ecosystem. However, river management decisions that prioritized flood protection and navigation, along with other human activities, have turned the natural land-building process into land loss. Throughout the river systems feeding the Gulf, these activities have reduced the amount of sediment for land building; sediments have been significantly reduced from the Mississippi River.

For several years, federal, state and local groups have conducted significant work on sediment management. The task force has taken this previous work and expanded it to begin to create a Gulf-wide approach to sediment management. The extensive attention given to this issue is reflected in the level of detail and actions presented.

A sediment management approach referred to as "strategic use" is needed in the Gulf to address land loss through sustainable resource management and land rebuilding and restoration. This strategic use approach should incorporate, where practicable and ecologically appropriate, the following recommended actions:

- Maximize beneficial use of navigational dredged material, where practicable and ecologically acceptable, for effective and sustainable habitat restoration.
- Increase dedicated dredging of river and other sediment sources, such as permitted offshore sediment shoals, for use in habitat restoration projects.

- Implement river reintroduction projects (i.e., diversions) that mimic more natural hydrologic processes and provide freshwater and sediments to rebuild, restore and nourish areas where wetlands have been lost, and to help sustain areas where wetlands have been restored.

Strategic use of sediments will result in long-term benefits to the public by increasing protection from storm surge and sea-level rise, improving navigation, as well as increasing fish and shellfish productivity through restoration of sustainable coastal wetlands and barrier islands. Success with this approach requires process improvements, funding and policies that will advance environmentally beneficial projects based on science. In addition, to advance implementation, better information is needed on the availability of sediment resources for ecosystem restoration, as tools to effectively integrate diversions and dedicated dredging with other management activities, including navigation, flood damage risk reduction and water supply. Tools are also needed to coordinate and schedule beneficial use efforts and to ensure the most efficient and cost-effective transportation of materials for effective beneficial use.

Maximize Beneficial Use Where Practicable

On average in recent years, the U.S. Army Corps of Engineers and local partners contract for the dredging of approximately 100 million cubic yards of sediments annually from federal navigation channels in the Gulf states.[82] Due to such things as funding limitations, only a small percentage of all material dredged from federal channels in the five Gulf states is currently used beneficially, while the remainder is disposed of in open water or designated disposal facilities. While not all dredged material may be the right consistency or composition to be used beneficially for ecosystem restoration, a large amount of sediment is available as a resource (but not currently used) for effective beneficial use in ecosystem restoration. The time required for implementation of a beneficial use project can vary considerably, from one to three years or longer. It depends upon the time of year, the amount of material being dredged, the distance the material needs to be transported for disposal, the nature of the site to be dredged and of the placement site, weather conditions, and whether the dredging was an emergency response to keep navigation channels open. The opportunities for beneficial use are often missed as a result of such things as funding limitations and lack of coordinated planning. Beneficial use can appear to be a more costly option up front than traditional disposal methods; however, if ecological benefits were factored into cost estimates, beneficial use could be more comparable.

Recommended actions to maximize, where practicable, beneficial use projects:

- Commit to using existing authorities and funding sources in law and policy and improving collaboration among federal, state, tribal and local entities to ensure the maximum beneficial use of sediment, including maintenance-dredged material, where effective, practicable and protective of aquatic life.
- Identify opportunities for immediately improving beneficial use of dredged material by overlaying dredging schedules with restoration project construction.
- Evaluate the potential for the creation of a network of permanent pipeline sediment delivery systems to transport material from dredges to coastal restoration projects.
- Develop options for funding the incremental cost of beneficial use of dredged material, including pooling public and private funds for these activities.

- Pursue implementation of authorized projects to maximize beneficial use where practicable.

Increase Dedicated Dredging

Unlike beneficial use projects, which derive sediments from the routine maintenance of navigation channels, dedicated dredging projects use sediments from ecologically appropriate sources to restore wetlands, barrier islands and shorelines. In sediment-starved coastal areas, the material removed in the course of dredging from river, permitted offshore shoals, and other sources often referred to as "external" sources can be used to restore critical habitats. Additionally, dedicated dredging can help to increase the amount of sediments in the coastal system, which aids in habitat sustainability and resiliency. Dedicated dredging has proven to be an efficient and effective near-term habitat restoration technique. To yield long-term results, dedicated dredging or beneficial use may be coupled with river reintroduction projects to help ensure the sustainability of the restored areas.

Recommended actions to increase dedicated dredging projects include:

- Use a sediment management approach to strategically target project selection and incrementally implement projects using a regional sediment management approach. The approach should determine the sediment budget, identify ecologically appropriate alternatives, and then place sediment trapping and sediment transport infrastructure to reduce restoration project costs over time. This approach should also consider any potential benefits to the navigation system of removing sediments from rivers for the purpose of habitat restoration, such as the reduced need for navigational maintenance dredging downstream.
- Identify potential sources of public and private funds to help implement dedicated dredging projects for habitat restoration. This effort should include exploring currently available, but potentially underutilized, authorities and funding sources.

Implement River Reintroduction (Diversion) Projects

Use of diversions, or river reintroduction projects, is a technique to restore altered hydrologic patterns by conveying suspended sediments and freshwater from rivers to mimic more natural river delta processes and thereby rebuild land and restore wetlands. River reintroduction projects are important tools for habitat restoration. To yield long-term results, river reintroduction (diversion) projects may be coupled with dedicated dredging projects to help ensure sustainability of restored areas.

Recommended actions to implement river reintroduction projects:

- Expedite construction of river reintroduction projects that have been authorized, planned and designed.
- Develop and implement necessary scientific protocols and monitoring to determine the long-term effectiveness of diversion projects.
- Develop procedures to make diversion project implementation proceed more efficiently, including processes for establishing project partnership agreements between federal and state agencies.

- Identify potential sources of public and private funds, and consider opportunities for increased flexibility to help expedite implementation of reintroduction projects for habitat restoration.
- Explore the development of alternative restoration project delivery approaches that could effectively leverage state, local and private sector partnerships.

Restore and preserve more natural river processes of sediment and freshwater distribution

Inland and coastal wetlands, estuaries, barrier islands and natural beaches and dunes provide critical ecosystem services. These habitats act as storm buffers, foster sustainable commercial and recreational fisheries and other important natural resources, provide habitat for the recovery of threatened and endangered species, protect important cultural resources, and support tourism and other recreational economies and jobs. These functions can be protected and enhanced by reconnecting rivers with their deltaic plains and managing sediment and freshwater inflows while still allowing flood damage risk reduction, navigation and other economic activities to inform appropriate management actions. A comprehensive approach would ensure that these areas are managed for the maximum benefit for the entire ecosystem. Reintroducing natural deltaic processes that restore sediment to vulnerable areas would help ensure the sustainability of coastal ecosystems.

In the Gulf, communities, commerce and ports have long relied on the extensive network of human-made flood damage risk reduction and navigation structures for their existence. River management priorities historically centered on navigation and flood risk reduction. While successful in meeting those two goals, those priorities created unintended consequences to the surrounding environment by accelerating wetland and barrier island erosion and restricting the flow of vital sediments that had sustained the ecosystem over time. Oil and gas industry canals, pipelines and other infrastructure crisscross the landscape to accommodate exploration, development and commercial activity related to these enterprises, further accelerating coastal land loss.

As a consequence of these and other stresses, Louisiana alone currently experiences about 90 percent of the total coastal marsh loss in the continental United [83] Current U.S. Geological Survey (USGS) analyses show that coastal Louisiana has lost 1,883 square miles of land from 1932 to 2010.[84] Between 1985 and 2010, the average rate of land loss was 16.57 square miles per year.[85] If this loss were to occur at a constant rate, it would equate to Louisiana losing a wetland area the size of one football held each hour.[86]

Recommended actions include:

- Expedite the Louisiana Coastal Authority's Mississippi River Hydrodynamic and Delta Management Studies as outlined in the Water Resources Development Act of 2007. These studies, led by the U.S. Army Corps of Engineers and the State of Louisiana, would investigate strategies to take maximum advantage of the water and sediments of a dynamic Mississippi River system for coastal restoration, while also maintaining economically viable navigation and providing storm damage risk reduction. The studies would also identify stakeholder interests, potential outcome targets, and a process for developing decision-support tools. Also, these studies should include a technically thorough exploration of large-scale riverine management strategies and should be informed by the State of Louisiana Master Plan, the

Mississippi Coastal Improvement Program, and other programs and plans. This important effort will look at the future needs of the coastal ecosystem with those of navigation, future land use and management, and the sustainability of coastal communities and economies.

Expand the network of state, federal and private conservation areas to ensure healthy landscapes that support the environment and culture of the region and the diverse services provided by the Gulf of Mexico ecosystem

Intact natural and restored ecological systems, watersheds, lands used for the production of food and other crops, recreational areas, and coastlines are essential to ensure the sustainability of ecosystem function and services. Habitat loss and fragmentation—the division of previously uninterrupted habitat—threaten the ecosystem's ability to support wildlife and recreation. Expanding the existing conservation base and enhancing ecological connectivity will provide greater resiliency to fish and wildlife habitat as well as to coastal communities. State and federal agency collaboration is an important step toward developing a Gulf-wide network of conservation areas to protect habitats and wildlife, support ecosystem services, and ensure recreational and commercial opportunities. This will require improved alignment of federal, state and other land conservation efforts and funding to identify and protect key conservation areas and habitat corridors. A coordinated, stakeholder-driven, science-based effort is needed to identify priority habitats and linkages to serve as core areas, buffers and potential corridors.

State and federal agencies and NGOs are implementing ecosystem-scale planning and protection of important Gulf Coast habitats. For example, Louisiana's existing Coastal Forest Initiative, established to acquire coastal forests including cypress and tupelo from willing sellers, protects a highly significant habitat type while providing natural storm surge buffers. This initiative is the type of conservation effort that should be supported and expanded. Building on these types of efforts will reinforce Gulf-wide actions to restore, conserve and protect important coastal habitats and landscapes.

Recommended actions include:

- Improve collaboration across federal, state, tribal, local and private lands to form habitat corridors of connected areas for key species in mutually designated priority areas.
- Expand collaboration with stakeholders to develop a conservation framework for appropriate protection and compatible use.
- Solicit input regarding public use and conservation priorities in order to prioritize designation of sites.
- Bring together the Land and Water Conservation Fund agencies (DOI, NOAA and USDA), as well as private landowners, private partners, states and other relevant federal agencies, to achieve high-priority landscape-scale conservation projects.

Restore and conserve coastal and near-shore habitats, with a focus on marshes, mangroves, seagrasses, barrier islands, natural beaches and dunes and coastal forests and prairies

A variety of coastal habitat types defines the Gulf of Mexico. These areas provide nurseries, food and habitat for numerous species of commercially and recreationally important finfish and shellfish, as well as migratory birds and a diverse array of mammals, amphibians and reptiles.

- *Fresh, intermediate, brackish and salt marshes and mangroves* trap and hold sediments, improve water quality by filtering input from rivers and runoff from adjacent uplands, and serve as natural coastal buffers to protect shorelines.
- *Seagrasses* are a significant habitat across the Gulf Coast, providing a nursery function for recreationally and commercially important fish and wildlife, stabilizing the bottom, and serving as valuable recreational fishing grounds (see "Seagrasses in the Gulf").
- *Coastal forests and grasslands* such as bottomland hardwoods, longleaf pine, cypress-tupelo swamp and tall grass prairie assist with carbon sequestration and storm barrier protection, wildlife habitat and recreation. These habitats are integral to the hemispheric-scale neotropical bird migration.
- *Barrier islands serve* an important role, protecting mainland shorelines from storms and providing sandy beach and dune habitats and recreational areas.

SEAGRASSES IN THE GULF

Healthy seagrass beds are indicative of productive coastal and estuarine ecosystems, providing habitat and forage for waterfowl, fish, protected species and shellfish; buffering against storms; stabilizing sediments; and improving water quality. However, these important shallow water marine habitats are in trouble, with declines in seagrass acreage ranging from 4 percent to 95 percent in bays and estuaries of the Gulf of Mexico.

Significant declines in seagrasses have been documented for Galveston Bay, coastal Louisiana, the Mississippi Sound, Mobile Bay and most of the bay systems of the Florida Gulf coast, with the coastal bend and Laguna Madre of Texas remaining relatively stable. Not only are seagrasses disappearing, but they are also changing in species composition, densities and patchiness. The causes of loss and changes of seagrasses are many, including tropical storm events, nutrient loading, dredging impacts and direct physical disturbances such as waterfront construction, bottom trawling and propeller scarring.

Source: U.S. Geological Survey. 2004. Seagrass habitat in the Gulf of Mexico: Degradation, conservation and restoration of a valuable resource. USGS 855-R-04-001. <http://gulfsci.usgs.gov/gom_ims/pdf/pubs_gom.pdf>

Overharvesting, development, pollution and a range of other factors have sped the loss of these important habitat types. Restoration efforts should focus on addressing these losses. Water quality protection, living coastal and marine resource conservation, and increasing the pace of direct restoration efforts—including incorporating ecosystem restoration into shoreline stabilization efforts through the use of "living shorelines" and similar approaches—are needed to address these issues.

Recommended actions include:

- Develop an inventory (types, locations, status and other criteria) and tools to set priorities for restoration investments for seagrasses, mangroves, coastal forests and marshes. This inventory will be critical to developing a strategic Gulf-wide approach and assessing its results.
- Provide engineering and design assistance to local governments and private homeowners for creating living shorelines and fish and wildlife habitat.
- Coordinate efforts by private landowners, partner organizations, and local, state, tribal and federal governments to connect conservation activities on coastal lands. This effort will help establish buffers landward of marsh shorelines which may reduce the effects of sea-level rise.
- Implement protective measures to reduce propeller scarring and other physical damage to seagrass beds.
- Increase the acreage of watershed, riparian, wetland and shoreline habitats through conservation, restoration and protection actions to maximize water quality, recreation, and fish and wildlife benefits.

Restore Water Quality

Water quality is a significant indicator of the health of the Gulf of Mexico. The condition of the Gulf's waters reflects alterations in natural hydrology and pollution from urban development, industry, agricultural runoff, atmospheric deposition, and other sources throughout the entire Gulf watershed.

One of the most striking examples of water quality impacts in the Gulf can be found stretching from the mouth of the Mississippi River to the upper Texas coast. There, each summer, a hypoxic (low-oxygen) area or "dead zone" forms, primarily caused by excess nutrients in the water, which deplete oxygen that organisms need to survive. A significant amount of these nutrients (mostly nitrogen and phosphorus) comes from fertilizers, soil erosion, atmospheric deposition, and discharges from wastewater treatment plants and water control structures. These nutrients are carried to the Gulf of Mexico from throughout the entire watershed and upper basin states via the Mississippi and Atchafalaya Rivers. Nitrate loadings in the Mississippi River rose from 200,000 to 500,000 tons per year in the 1950s and 1960s to an average of about 1,000,000 tons per year during the 1980s and 1990s.[87]

Hypoxia in the Gulf of Mexico is an obstacle to achieving healthy, diverse and sustainable fisheries because it creates an uninhabitable environment for marine life. Hypoxia is most detrimental to benthic (bottom-dwelling) organisms such as crabs, oysters and other species in early life stages. These organisms, unable to migrate away from the zone, ultimately die in the low-oxygen conditions. The annual loss of these benthic species affects the entire food web and also has implications for the Gulf economy.

Pathogens, nutrients and other pollutants carried by stormwater and wastewater, as well as HABs (or "red tide") carried by ocean currents, have significant impacts on Gulf water quality, affecting the Gulf's ability to maintain healthy populations of aquatic organisms and healthy recreational waters. Urbanization alters the ecosystem's capacity to absorb runoff, leading to higher stormwater flows with increased flooding and discharge of pollution to receiving waters. Runoff from agricultural operations also contributes to excess sediment and nutrient levels in Gulf waters. Combined, these stressors degrade drinking water sources and recreational waters and commercial fisheries, cause closings of beaches and shellfish-growing areas, and reduce the quality of aquatic habitat and human health.

In addition to inputs of nutrients and pollutants from coastal systems and watersheds, deposition of air emissions also contributes to water quality impacts, particularly in near-coastal waters. Nitrogen and other airborne pollutants, such as sulfur in the form of sulfuric acid, alter surface seawater alkalinity, pH and inorganic carbon storage. These effects disrupt natural biogeochemical cycles.[88]

Across the Gulf region, freshwater flows need to be restored to more natural conditions. Rivers and streams provide freshwater inputs that help to maintain salinity gradients and are the source of nutrient and sediment inputs that, in proper combination, produce ecologically sound and healthy estuaries. Hydrologic modifications have affected estuaries throughout the Gulf by altering the amount of freshwater delivered. These estuaries depend on freshwater inflow to sustain their fisheries resources, particularly oysters, as well as to support habitats such as seagrass meadows, near-shore reefs, coastal marshes and mangroves.

Water quality in the Gulf can be most effectively improved by focusing on the principal sources of water quality degradation. Major actions to restore water quality in the Gulf of Mexico are described below.

Major Actions

Decrease and manage excess nutrient levels in the Gulf through the development and implementation of state nutrient reduction frameworks

Supporting the development and implementation of state-developed nutrient reduction frameworks in the Gulf and the Mississippi River Basin states will be important for reducing HABs and hypoxic conditions in the Gulf, such as the "dead zone," and improving local water quality conditions.

Beginning with the release of its first Action Plan in 2001, and reaffirmed in the updated 2008 Action Plan, the Hypoxia task force established a collaborative interim goal to reduce the five-year running average areal extent of the Gulf of Mexico hypoxic zone to less than 5,000 square kilometers (1,931 square miles).[89] As of the most recent survey (2011), the current five-year average is 17,350 square kilometers, or 6,700 square miles[90] (see "Area of Mid-Summer Bottom Water Hypoxia").

To accelerate progress on and effectiveness of this goal, the task force recommends working in cooperation with the Hypoxia Task Force and the Mississippi River Basin states to support the development and implementation of state-developed nutrient reduction strategies. Specifically, the 2008 Action Plan states the objective as:

> Complete and implement comprehensive nitrogen and phosphorus reduction strategies for states within the Mississippi/Atchafalaya River Basin encompassing watersheds with significant contributions of nitrogen and phosphorus to the surface waters of the Mississippi/Atchafalaya River Basin, and ultimately to the Gulf of Mexico.[91]

In combination with this effort, Gulf states not part of the Hypoxia Task Force (Alabama, Florida and Texas) are collaborating, through GOMA, to strengthen supporting science and technical capabilities and to develop their own state nutrient reduction frameworks to restore local water quality conditions.

Building on the states' water quality standards programs, key elements recommended for effective state nutrient reduction frameworks include:

- Prioritizing watersheds on a statewide basis for nitrogen and phosphorus loading reductions.
- Setting watershed load reduction goals based on the best available water quality information.
- Ensuring the effectiveness of point source permits in targeted/priority subwatersheds.
- Developing more effective reduction measures for nonpoint sources and other point sources of storm water not designated for regulation.
- Partnering with federal and state agencies, NGOs, the private sector, landowners, and other agriculture partners to develop watershed-scale plans and promote adoption of science-based nutrient management conservation practice systems that offer enhanced environmental protection, and may also increase agricultural production.
- Identifying stormwater and septic system tools.
- Establishing accountability and verification measures.
- Conducting annual reporting on load reductions and impacts in targeted watersheds.

Successful implementation of these frameworks could significantly advance the Strategy's goal to restore water quality as it relates in particular to reducing excess nutrients to Gulf of Mexico coastal waters. Consequently, the task force, through its coordination - responsibilities with the Hypoxia Task Force, should explore means of providing technical and resource support to this effort.

Recommended actions include:

- Accelerate the development and implementation of the Hypoxia Task Force's state-developed nutrient reduction frameworks by the states in the Mississippi Atchafalaya River Basin.
- Coordinate the task force's efforts with the Hypoxia Task Force's next reassessment, scheduled to be completed in 2013. This coordination will help better ensure consistency and focus related to the establishment of performance indicators and supporting actions for the reduction of nutrients.
- Promote the development and implementation of state nutrient reduction frameworks by Gulf states not part of the Hypoxia Task Force (Alabama, Florida and Texas).
- Provide technical assistance and explore additional ways to expand support for implementation of state nutrient reduction frameworks. This includes supporting state-led management efforts such as installation and monitoring of best management practices (BMPs) aimed at reducing excess nutrients flowing into the Gulf.

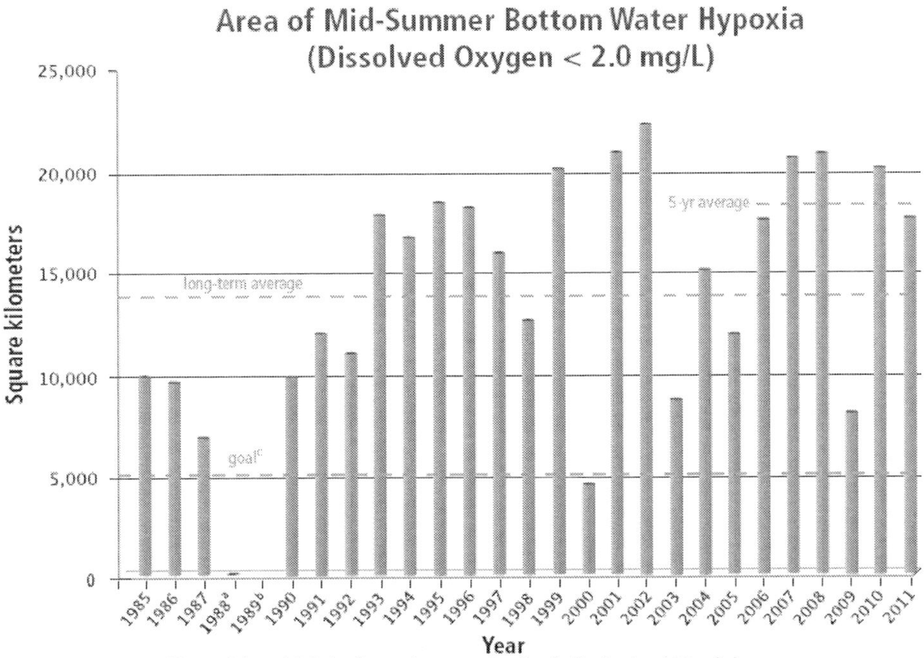

[a] Only 40 sq km were affected in 1988 (a drought year in the Mississippi Basin).

[b] No data were collected in 1989.

[c] Mississippi River/Gulf of Mexico Nutrient Task Force Hypoxia Action Plan goal.

Data source: N.N. Rabalais, Louisiana Universities Marine Consortium, R.E. Turner, Louisiana State UniversityFunded by: NOAA, Center for Sponsored Coastal Ocean Research.

MISSISSIPPI RIVER BASIN INITIATIVE

To improve the health of the Mississippi River Basin, USDA's Natural Resources Conservation Service (NRCS) has established the MRBI. Through this initiative, NRCS and its partners help producers in selected watersheds in the Mississippi River Basin voluntarily implement conservation practices that avoid, control and trap nutrient runoff; improve wildlife habitat; and maintain agricultural productivity.

NRCS dedicated more than $70 million in fiscal years 2010 and 2011 combined to fund conservation projects to improve water quality in priority watersheds throughout the Mississippi River Basin.

Focus restoration actions in priority watersheds to address excess nutrients in coastal waters and reduce hypoxic conditions

One of the most cost-effective and efficient ways to reduce sources of excess nutrients is to target program funding and technical assistance for accelerated treatment of areas with the most critical need and in the highest-priority watersheds. This approach, which addresses local water quality and natural resource concerns at a small watershed scale, will contribute to reducing runoff of excess nitrogen and phosphorus on a large scale, and contribute to reducing hypoxia and HABs in the Gulf. As noted above, state nutrient reduction frameworks are useful mechanisms for identifying priority watersheds.

Recommended actions include:

- Using a science-based approach, in collaboration with federal and state agencies, select and target resources for priority sub-watersheds where significant opportunities exist to reduce excess nutrients, sediments and pathogens flowing into the Gulf of Mexico and its estuaries.
- Increase and coordinate conservation practices on agricultural lands to enhance water quality in priority watersheds within the Mississippi River basin and other tributaries in the Gulf states through USDA's Mississippi River Basin Healthy Watersheds Initiative (MRBI) and similar programs (see "Mississippi River Basin Initiative").
- Broaden implementation support for the recommendations in the Conservation Effects Assessment Program (CEAP) in the upper Mississippi River (see "Conservation Effects Assessment Program").
- Increase the number of Comprehensive Nutrient Management Plans (CNMPs) implemented for animal feeding operations and pasture grazing. In conjunction with the CNMPs, implement necessary complementary conservation and BMPs including fencing, hard-bottom cattle crossings, constructed wetlands and other BMPs most appropriate for the individual location and operation to effectively address nutrient transport from agricultural lands. This action would also help to reduce pathogens in the Gulf of Mexico.
- Coordinate among state and federal programs that promote riparian buffer restoration and preservation to better align and maximize efforts.
- Work with agricultural equipment and fertilizer dealers to make precision agriculture equipment and technologies available to private landowners and operators at affordable prices.

Reduce pollutants and pathogens from stormwater flows and other sources

Contaminants carried within stormwater and sewer overflows have the potential to affect both aquatic life and human health. These contaminants can include heavy metals from urban infrastructure like roads, guardrails, and construction materials; fertilizers and pesticides from residential and commercial use; petroleum products and paving compounds such as polycyclic aromatic hydrocarbons (PAHs) from commercial and residential sources; bacteria, viruses, protozoa and parasites from fecal waste; and nitrogen and phosphorus pollution. In the Gulf, these contaminants can impact recreational waters and commercial fisheries, as well as the overall quality of aquatic life. The task force supports a region-wide effort to reduce pollutants and pathogens stemming from stormwater flows and other sources entering Gulf Coast waters.

Recommended actions include:

- Promote comprehensive solutions that may include green infrastructure and low-impact development approaches in urban and suburban areas to help reduce combined sewer and sanitary sewer overflows, untreated stormwater runoff, as well as producing natural habitat buffering benefits.
- Facilitate the expansion of municipal stormwater permit programs to include fast-growing suburbs and urban areas not currently covered by these programs.

- Encourage states to restrict phosphorus in lawn fertilizer, following the successful examples of Maine, Maryland, Michigan, Minnesota, New Jersey, New York, Virginia, Washington and Wisconsin.

CONSERVATION EFFECTS ASSESSMENT PROJECT

CEAP is a multi-agency, multi-resource effort led by NRCS to assess the effects of conservation practices on the nation's cropland, grazing lands, wetlands, wildlife and watersheds.

A recent CEAP study, *Assessment of the Effects of Conservation Practices on Cultivated Cropland in the Upper Mississippi Basin*,[92] found that farmers have adopted conservation practices resulting in significant reductions in nutrient losses. However, the study also found that conservation plans that include comprehensive nutrient management are generally lacking throughout the region. According to the report, a suite of practices that includes both soil erosion control and comprehensive nutrient management is needed to simultaneously address soil erosion and nitrogen leaching loss.

Improve the quality and quantity of freshwater flow into priority estuaries to protect their health and resiliency

Estuaries, transition zones between the coastal watersheds and offshore marine habitats, serve as barometers of the health of the Gulf of Mexico. They provide breeding grounds, forage and nursery areas for recreational and commercially important species of fish and wildlife; habitat for benthic species such as oysters and bay scallops; and transitional and migratory wildlife corridors between upstream freshwater habitats and offshore marine waters.

Gulf estuaries are affected by multiple stressors, including nutrient loading, other pollution, altered hydrology resulting in unnatural tidal exchange and/or salinity gradients, and upstream dams. While levees are sometimes necessary to enable management of unnatural systems in a manner that mimics nature, barriers to natural tidal flow such as weirs, levees, causeways and artificial canals can cause habitat degradation and loss by disrupting the natural salinity variations necessary to support a diversity of coastal organisms (e.g., oysters, saltmarshes, seagrasses). Moreover, as demand for freshwater resources continues to increase throughout the Gulf Coast river basins and underground aquifers, maintaining sufficient freshwater flow into the bays and estuaries will become increasingly challenging.

Recommended actions include:

- Support state and local government efforts to better protect groundwater supplies and instream freshwater flows by minimizing water loss from leaking water supply systems. Also, develop policy and incentives for water reuse on agricultural lands, urban areas (for watering lawns and golf courses), and other non-drinking uses of freshwater in areas where these resources are limited.
- Develop a science-based freshwater inflow regime and monitoring system to improve the timing, quantity and quality of freshwater inflow within managed river systems. This approach would not only benefit wetlands, but also remove nutrients from direct

introduction to Gulf waters and reduce hypoxia, and sustain important fisheries, oysters and protected wildlife.

- Identify and develop means and opportunities to improve water management planning, conservation, land-use planning, and barrier evaluations (to tidal flow) to restore and maintain freshwater resources and estuaries along the Gulf. Consider removal, addition or modification of tidal barriers where cost-effective scientifically supported solutions can be shown to improve and restore estuarine habitat.

Coordinate and expand existing water quality monitoring efforts supporting adaptive management of programs and projects designed to improve water quality

Available water quality data throughout the Gulf of Mexico basin (from inland watersheds to estuaries/nearshore areas and the offshore waters) are insufficient or inadequate as currently collected to allow for the accurate assessment of status and trends. Incomplete data make it difficult to quantify changes and to determine if restoration and protection measures are successful. The task force supports the following actions to better coordinate and expand water quality monitoring to support adaptive management:

- In collaboration with federal and state partners, assess current water quality monitoring information and data. Identify gaps in core information needs and performance measures to develop a long-term water quality monitoring program building on existing efforts, including assessment of nutrients, HABs and other water quality parameters.
- Monitor and assess nutrients in the Mississippi River Basin, to better quantify current nutrient loading to the Gulf and target reduction efforts.
- Leverage existing information management infrastructures of EPA, NOAA, USDA and DOI to integrate and provide monitoring data and information supporting local, state and regional decision needs.

Collaborate with Mexico to assess and reduce emissions from oceangoing vessels in the Gulf that degrade water quality

Water quality, particularly in near-coastal waters, can be negatively affected by pollutants emitted from oceangoing vessels. To protect marine and coastal areas, including the Gulf, the United States and Canada adopted the North American Emissions Control Area (ECA) in March 2010. This initiative will put in place lower sulfur marine fuel standards for oceangoing ships operating in the ECA beginning in August 2012, as well as nitrogen oxide standards for engines on ships built in 2016 and later.[93] It will dramatically reduce air pollution from ships and deliver water quality benefits by reducing atmospheric deposition of pollutants to coastal ecosystems.

Ongoing efforts will continue with Mexico to increase awareness of the health and environmental benefits of reducing emissions from oceangoing vessels. Data from a recent U.S.-Mexico fuel switching demonstration, for example, show that particulates and sulfur dioxide emissions from container ships can be reduced by up to 80 percent and 89 percent, respectively.[94] This plan presents a unique opportunity to increase the benefits of reducing ship emissions in the Gulf of Mexico.

Replenish and Protect Living Coastal and Marine Resources

The Gulf Coast is inhabited by a rich diversity of mammals, birds, reptiles, fish, invertebrates and plants. For example, the coral reef systems support some of the most diverse assemblages of life, including anemones, spiny lobsters, yellowtail snapper and nurse sharks. These species provide the underpinning for economically critical commercial and recreational opportunities, as well as many vital ecosystem services such as filtering pollutants, sequestering carbon, and contributing to a robust and biologically diverse system better capable of rebounding from natural and human-made events. Gulf of Mexico habitats are among the most important on the continent for wintering waterfowl and migrating songbirds. Radar images show hundreds of millions of migrating birds passing through the Gulf each spring and fall, relying on coastal habitats for rest and food. Beach-nesting birds and marsh birds are in decline and other beach-dwelling species are at increasing risk due to threats such as climate change and habitat loss. Anadromous fish such as striped bass, alligator gar and the threatened Gulf sturgeon need healthy, connected coastal waterways and rivers flowing into the Gulf for spawning.

Human communities depend substantially on the productivity of living coastal and marine resources for economic and recreational benefit:

- A 2006 national survey of wildlife-related recreation reported that revenues from fishing, hunting and wildlife viewing in the Gulf region topped $22 billion.[95] In 2006, almost 1.6 million people participated in wildlife viewing in Florida alone. Of those people, about 1.3 million viewed shorebirds, 800,000 viewed fish and 700,000 viewed ocean mammals.[96]
- Recreational and commercial fishing is a multi-billion-dollar industry critical to the economies of each of the Gulf states. From 2007 to 2009, more than 75 percent of the total U.S. shrimp landings and over 60 percent of oyster landings were from the Gulf of Mexico.[97] More than 44 percent of all marine fish caught by recreational anglers in the United States in 2009 were from the Gulf of Mexico.[98]

Many Gulf Coast plant and animal species are in decline or at abundances far below historical levels. Among the threats contributing to the depletion of living coastal and marine resources are development in environmentally sensitive areas, land loss, predation, habitat disturbance, pollution, water quality degradation (including excess nutrient discharge leading to hypoxia), and increases in disease and invasive species.

Implementing the recommendations provided in the "Restore and Conserve Habitat" and "Restore Water Quality" sections of this Strategy will directly and significantly benefit and improve living coastal and marine resources. The sustainability of commercial fish species, for example, depends on healthy coastal habitat; nearly 97 percent of commercial finfish and shellfish landed in the Gulf of Mexico depend on estuaries and coastal waters at some point in their life cycle.[99] The survival and health of estuaries and offshore ecosystems are linked with the riverine processes that deliver freshwater, sediment and nutrients to moderate salinity, sustain floodplain habitats and enhance coastal fisheries. The interconnectedness and complexity of near-shore and offshore habitats create opportunities for multiple benefits from ecosystem restoration actions. Successful restoration of living coastal and marine resources will entail a robust monitoring effort, implementation of species restoration plans, and targeted reintroduction and re-stocking of depleted resources.

Major Actions

Restore depleted populations of living coastal and marine resources

The Gulf is home to a number of endangered and threatened species at risk of extinction. Twenty-eight species listed under the Endangered Species Act occur in the Gulf—three plants, four birds, nine mammals, seven reptiles, three fish and two corals.[100] Additionally, many species of marine mammals in the Gulf are protected under the Marine Mammal Protection Act. Numerous other species are listed in state wildlife protection programs. Recent management efforts of federal and state fisheries have reduced the number of fish stocks on current "overfishing" and "overfished" lists; however, commercially and recreationally important species have been reduced from historical levels. For example, red snapper in the Gulf have been reduced to less than 10 to 15 percent of historic spawning abundance,[101] though recent regulatory accountability measures under the Magnuson-Stevens Fisheries Conservation and Management Act are making significant strides to rebuild this stock.[102]

Restoration actions that improve habitat and restore water quality will substantially improve the health of the living coastal and marine resources of the Gulf. However, some depleted populations need additional management assistance to fully recover and become self-sustaining components of the ecosystem. For example, nesting populations of Kemp's Ridley sea turtle have experienced significant rebuilding as result of the management, restoration and conservation efforts of Mexico and the United States.[103] Additionally, 2011 marked the historic reintroduction of 10 endangered whooping cranes, after a 40-year absence, into the marshes of southwest Louisiana.[104] Restoring reefs in areas of significant vessel groundings and oyster reefs in once-degraded habitat is another example of techniques to replenish living resources.

Recommended actions include:

- Revisions to existing management plans—e.g., Endangered Species Act (ESA) Recovery Plans, Magnuson-Stevens Act Fishery Management Plans, or State Wildlife Action Plans—should use an ecosystem-based approach to improve monitoring and research of living coastal and marine resources, including nursery, spawning and breeding habitats; integrate input from all affected stakeholders; minimize adverse impacts that can result from human activities; maintain sustainable native populations; and protect, restore and maintain critical habitat for fish, wildlife and plants in coastal areas.
- Implement existing plans and, where appropriate, develop plans for conserving threatened and endangered species. Most species listed under the ESA have recovery plans that prioritize actions needed for recovery. Efforts should be made to implement high-priority recovery actions and complete recovery plans for remaining

threatened and endangered species currently not addressed in an approved recovery plan.

- Improve the quality and timeliness of fishery-dependent data (e.g., from recreational catches) and implement a Gulf-wide, comprehensive and long-term fishery-independent data collection program (e.g., from scientific research cruises). Decrease the time between data collection and its use for fishery management decisions. Lacking or lagging data are frequently cited as a major challenge in achieving sustainability and maximizing economic benefits to recreational and commercial fisheries.
- Explore opportunities to enhance depleted fishery stocks through reintroductions, and/or develop and expand capabilities to re-stock native species of fish, shellfish, plants and wildlife.
- Identify beach-nesting bird colonies (e.g., least terns, black skimmers) and solitary nesting species (e.g., piping plover, American oystercatcher).
- Investigate the impacts of mechanical disturbance of subsurface habitat on the sustainability of living coastal and marine resources.

Conserve and protect offshore environments

One hundred fifty million years of biological and geological history have created numerous banks, escarpments and hard grounds that ring the continental margins of the Gulf of Mexico. These hardbottom features are relatively rare, highlighting the importance of conserving these habitats where they do exist. The seafloor environments formed by these features are home to dense communities of corals, sponges and other invertebrates. They also attract resident and migratory fish species, including snapper, grouper, sharks and rays, as well as marine mammals and sea turtles. These ecologically important places extend into deep areas of the Gulf where dense assemblages of deepwater coral species and seeps of methane support unique and biologically rich marine communities. The deepwater environments of the Gulf support schools of squid, which are food sources for many whales, swordfish and marlin. Furthermore, the offshore environment serves as a critical component of the life cycles of many vertebrate and invertebrate species (e.g., white and brown shrimp, Gulf menhaden), so its health is integral to species' survival.

The unique hard-bottom structures found throughout the Gulf (such as the Florida and Texas Snapper Banks, the Alabama Pinnacles, the Alabama Alps, the Florida Middle Grounds and Pulley Ridge, and the Texas and Louisiana Flower Garden Banks) support prolific invertebrate communities and important reef fish spawning areas. Pulley Ridge is a series of drowned barrier islands on the southwest Florida Shelf, which are the deepest known light-dependent coral reefs in the United States. Throughout the Gulf, extensive offshore mats of *Sargassum* provide shelter, nursery habitat and food for deepwater species, including sea turtles. These highly diverse concentrations of marine life are essential to maintaining the Gulf's biodiversity and the productivity of its commercial and recreational fisheries.

The health of these environments, and of all the coastal and offshore habitats of the Gulf of Mexico, relies on the physical and biological connections that link the places and their communities to each other and to the wider Gulf and Caribbean region. Therefore, protecting and managing a network of ecologically significant offshore sites will be important to the Gulf's overall biological productivity and resilience.

100-1000 Restore Coastal Alabama Project

100-1000 Restore Coastal Alabama is a collaborative partnership and volunteer effort to restore 100 miles of oyster reefs to help create the conditions needed to plant, support and promote more than 1,000 acres of coastal marsh and seagrass beds. This restoration effort will provide habitat for oyster larvae to settle and colonize, establish nursery habitat for commercially and recreationally important finfish and shellfish, dampen wave energy, decrease erosion, stabilize sediments, and improve water clarity. This coastal restoration project also includes critical job creation and community involvement components to support and sustain the vision of a better coastal Alabama.

Source: 100-1000 Restore Coastal Alabama. <http://100-1000.org>

Recommended actions include:

- Identify candidate locations for cooperative conservation and restoration of offshore living resources based on biological significance through focused mapping and exploration efforts.
- Use existing conservation frameworks and legal authorities, and develop prioritization criteria for appropriate protection and compatible use that accommodate both economic importance and biological integrity. This effort should be undertaken in collaboration with relevant stakeholders and build on existing efforts.

Restore and protect oyster and coral reefs, and other coastal environments

Oysters have significant economic, environmental and social value throughout the Gulf coast and are considered one of the important economic engines in the region, particularly in rural coastal communities. Coral reefs are among the most diverse and biologically complex ecosystems in the Gulf, supporting high levels of biodiversity and densities of living organisms, as well as providing high regional economic value through ecotourism. Oyster and coral reefs in the Gulf serve many ecological functions, including reducing shoreline erosion and storm surge, creating habitat for other species, and improving water quality by filtering nutrients and sediments. These habitats have declined as a result of altered salinities and sediment, degraded water quality, overfishing, disease, climate change, storms and droughts.

One important opportunity for enhancing coastal resiliency is through restoration of native reefs and other habitats such as salt marsh, mangroves, barrier islands, seagrass beds and other near-shore adapted plants and sessile organisms. This action can create a "living shoreline," which can be a viable alternative to an armored shoreline. These structures also provide the added benefits of improved water quality, and overall increase in habitat diversity and productivity. For example, oyster shell/coral/limestone breakwaters, if properly constructed as part of living shoreline projects, can serve ecological functions similar to those of natural reefs.

Recommended actions include:

- Develop criteria to prioritize oyster reef and coral reef creation, restoration and enhancement projects across the Gulf of Mexico.
- Engage aquaculture industry in oyster and other shellfish restoration efforts, building on successes in other regions of the country. Consider artificial reefs in territorial and coastal waters where appropriate, following best practices to minimize impacts on natural habitats and existing plant and animal communities.
- Restore and manage coral and oyster reefs to support a variety of ecosystem services, including sustainable harvest, fish production, water filtration, nitrogen removal, and shoreline protection and stabilization.

Coordinate and expand existing Gulf monitoring efforts to track sentinel species and sites

Sentinel species are representative plants and animals that can be tracked over time to help alert researchers, decision-makers and the public about current or potential threats, trends and impacts to the ecosystem. Sentinel species are often the first or most obvious species to show effects of environmental change, serving as the proverbial "canary in a coal mine." Sentinel sites are key locations in coastal and marine environments suitable for intensive study and sustained observations to detect and understand physical, chemical and biological changes in the ecosystems they represent. By focusing on sentinel species and sites, scientists can better understand the status and trends in ecosystem health, improve ecosystem risk models, target restoration actions, and better inform management actions.

Example Sentinel Species and Sites for Monitoring Gulf Ecosystem Health

Sentinel species could include:

- **Fish:** Atlantic bluefin tuna, red snapper, gag grouper, menhaden, whale sharks, Gulf sturgeon, flounder, speckled trout
- **Birds:** pelicans (brown and white), beach-dependent plovers (snowy, Wilson's, and piping), northern gannet (offshore), American oyster catcher, and colonial nesting birds (least tern, black skimmer, roseate spoonbill)
- **Mammals:** beach mice, sperm whale, manatee, dolphin
- **Turtles:** Kemp's Ridley, loggerhead, green sea turtles, diamondback terrapin
- **Corals:** staghorn, elkhorn, deepwater species
- **Plants:** spartina, mangroves (black and red), sea oats, turtle grass, *Sargassum*, algae
- **Invertebrates:** oysters, shrimp, bluecrab, macroinvertebrates

Sentinel sites could include:

- Deepwater coral reefs
- Fish spawning areas and other marine aggregating areas
- NOAA's National Marine Sanctuaries (NMS) National Estuarine Research Reserve Sites (NERRS) and EPA's National Estuary Program (NEP)

State and federal agencies in the Gulf are engaged in long-standing efforts to actively monitor fish and wildlife. These efforts provide the foundation for building sentinel programs. The DOI agencies, including the Fish and Wildlife Service (USFWS), the Bureau of Ocean Energy Management, the National Park Service, and USGS—as well as NOAA—all have monitoring programs that track federal trust species on and offshore. Additionally, EPA's NEP and NOAA's National Estuarine Research Reserves and National Marine Sanctuaries support extensive monitoring programs in the Gulf. Examples of long-standing surveys include the state-federal Southeast Area Monitoring and Assessment Program (SEAMAP), USFWS Waterfowl Population Surveys, Marine Mammal Population at-sea and aerial surveys, and bluefin tuna larval monitoring in the Gulf of Mexico. State agencies often have complementary monitoring programs. An accessible, comprehensive database of information is needed to coordinate and link efforts and inform management decisions.

- In collaboration with current partners, identify appropriate sentinel species and sites throughout the Gulf of Mexico that would provide long-term data to inform restoration actions.
- Support and enhance existing monitoring efforts to ensure that the selected suite of sentinel species and sites is sufficiently tracked. Pursue public-private partnerships to expand monitoring capabilities.
- Develop and apply methods to estimate population sizes; determine geographic distribution and abundance trends to implement protection strategies for sentinel species and sites of particular sensitivity.

Minimize, and eliminate where possible, invasive species that impact the Gulf of Mexico

Invasive species are species that are non-native to a specific ecosystem, or native species with a competitive advantage due to significant ecosystem impacts from human development, hurricanes or other factors. Invasive species can be plants, animals, or other organisms occupying terrestrial and/or aquatic habitats. More than 300 nonindigenous aquatic species alone have been found in the Gulf region, some of which are considered potentially harmful to the Gulf of Mexico ecosystem.[105] Some examples include nutria,[106] lionfish,[107] giant salvinia,[108] orange cup coral,[109] Asian tiger shrimp,[110] green mussel,[111] and several species of tilapia.[112] Invasive species are primarily introduced through human actions. For example, international shipping and aquaculture are major means of introduction of aquatic invasive species to the Gulf of Mexico.[113]

Because invasive species have few natural enemies and are often immune to native diseases, they usually spread rampantly. Also, altered habitat conditions from wetland loss and changes in stream flow enable invasive species to proliferate. Invasive species are one of the leading threats to biodiversity and have enormous consequences for agriculture, forestry, fisheries and other human activities, as well as human health. Control costs and environmental damages can add up to millions of dollars per year.[114]

Recommended actions include:

- Develop and implement Gulf states invasive species management plans and pathway-specific risk management plans. Use existing state and regional partnerships, such as the Florida Invasive Species Partnership and the Cooperative Invasive Species Management Areas, as potential model frameworks.

- Evaluate the impact of various introduction mechanisms (such as ship ballast water, aquarium and exotic pet trade, and the movement of large platforms into the Gulf) by collaborating with such entities as the Gulf of Mexico states, the Gulf and South Atlantic Regional Panel on Aquatic Invasive Species, the Aquatic Nuisance Species task force, the National Invasive Species Council and the Southeast Aquatic Resources Partnership.

Enhance Community Resilience

About 10 million people live along the Gulf Coast, enjoying the many unique benefits of living in this area. These communities also face challenges from natural disasters such as hurricanes and floods and from long-term hazards such as erosion, land subsidence and sea-level rise. Human-induced challenges also confront communities, including coastal land loss, population change and associated development, and technological disasters such as oil spills. These communities will benefit directly from Gulf restoration activities that will improve ecosystem services, such as better storm protection and healthier fisheries. Actions taken to increase community resilience will reduce future impacts from storms and sea-level rise, decreasing the duration and expense of response and recovery. In just the last 10 years, eight of the costliest hurricanes in American history caused an estimated $225 billion in damages to Gulf states.[115] In 2005, Congress made a supplemental appropriation of $94.8 billion to respond to the impacts of Hurricanes Katrina, Rita and Wilma.[116] A study commissioned by the Federal Emergency Management Agency found that every one dollar invested in proactive mitigation efforts to prevent damages yields four dollars in cost savings.[117] Some studies have shown even greater savings. Improving Gulf communities' capacity to address acute and chronic hazards will contribute to the economic viability of the region and the nation.

Each Gulf Coast community ultimately has its own needs, values and interests, so solutions for ecosystem restoration and coastal planning will be driven by local decisions based on local conditions. At the same time, to effectively meet immediate needs and anticipate future conditions, coastal communities—the decision-makers, residents, and other stakeholders—should collaborate better to identify solutions that consider Gulf-wide needs, regional concerns, and local conditions. Gulf states have initiated efforts to develop coastal improvement programs that address community resilience and ecosystem restoration in a manner that incorporates ecologically sustainable growth and local interests. These programs should be used to identify development priorities, such as maintaining important ecosystems, improving resilience of existing housing stock to better withstand disasters, or relocating housing and infrastructure out of sensitive areas to more appropriate, less-vulnerable locations. In order for communities to better plan, prepare and respond to changes associated with living along the coast, technical assistance and currently available analytical tools should be employed to increase resiliency of communities.

Major Actions

Develop and implement comprehensive, scientifically based and stakeholder-informed coastal improvement programs

Gulf Coast states should build upon and create effective coastal improvement programs to reflect state-based, stakeholder-driven priorities for coastal improvements that incorporate a range of federal and state coastal improvement and restoration programs.

The Mississippi Coastal Improvement Program (MsCIP), for example, is a framework for the protection, restoration, enhancement and re-establishment of more natural buffering capacities of coastal habitats. The State of Mississippi and the U.S. Army Corps of Engineers established this effort in response to Hurricane Katrina, and jointly manage the program. MsCIP addresses community resilience and ecosystem restoration in a manner responsive to natural system concerns and local interests. Although reducing risk was the major theme of the effort, the solutions presented in the plan inherently addressed land and habitat loss, living coastal resources and water quality.

SEAFOOD SAFETY

The *Deepwater Horizon* oil spill highlighted the value of fishery resources to coastal communities in the Gulf and the region's importance to the nation as a source of high-quality, abundant seafood. The Food and Drug Administration, NOAA's National Marine Fisheries Service, EPA, the U.S. Coast Guard, and the Gulf Coast states have taken unprecedented steps to verify the safety of the seafood harvested from the Gulf, first by closing areas exposed to the oil and then by establishing protocols designed to ensure the safety of seafood as fishing areas were reopened. All federal waters and state waters, with the exception of some small areas near the mouth of the Mississippi River, have been reopened.[118]

Many of the actions highlighted in this strategy will directly affect the quality and abundance of the seafood in the Gulf of Mexico, including addressing the loss of important coastal and marine habitats, improving water quality by reducing excess nutrients, and addressing sources of pathogens and other pollutants into coastal waters. In combination with the Gulf states' ongoing sampling and testing to verify the quality of fish, shrimp, crabs and oysters produced by their local industries, restoring habitat and water quality will contribute to a robust fishing economy in the Gulf of Mexico.

The MsCIP framework offers a model for developing local and state methods to achieve restoration at the local, state and regional ecosystem levels. The other coastal states have or are considering similar approaches. The State of Texas and the U.S. Army Corps of Engineers are pursuing a partnership to develop a comprehensive protection and restoration plan for the Texas coast based on existing authorities. Alabama has expressed interest in developing an effort similar to the MsCIP, and Florida is evaluating the benefits and potential approaches for pursuing a Gulf restoration plan. The State of Louisiana is focused on its State Master Plan update, which is scheduled to be completed in December 2012, and would look to further its partnership with the U.S. Army Corps of Engineers to create a common federal/state vision.

To effectively address ecosystem restoration, federal and state agencies need to coordinate comprehensive implementation efforts. They should identify shared priorities, maximize opportunities for efficiencies, provide scientific and technical support, and identify resources needed to assist in implementation.

EPA's Climate Ready Estuaries Program

EPA's Climate Ready Estuaries (CRE) is working to promote climate change adaptation with NEP partners in the Gulf of Mexico. Estuaries and coastal areas are particularly vulnerable to climate variability and change. To maintain water quality and protect coastal resources, managers may need to develop and implement adaptation measures. CRE provides resources to NEPs and other coastal managers to (1) assess climate change vulnerabilities, (2) develop and implement adaptation strategies, (3) engage stakeholders, and (4) share lessons learned with their peers and coastal communities.

CRE has supported several projects with Gulf of Mexico NEPs that have taken the lead in regional adaptation planning. Recent work has focused on developing strategies to enhance community resilience through vulnerability assessments and public engagement about sea-level rise. For example, in 2009 the Charlotte Harbor NEP partnered with the City of Punta Gorda, Florida, and finalized a climate change adaptation plan for the municipality. Currently, the Tampa Bay Estuary Program is leading a project with all of the Gulf of Mexico NEPs to develop a Gulf-wide resource guide on how to incorporate climate change into projects that restore coastal habitat. CRE shares lessons from demonstration projects it supports via a website (www.epa.gov/cre), annual progress reports, social media and lessons-learned summaries.

Recommended actions include:

- Advance efforts to create and/or implement coastal planning programs with a focus on Gulf-wide coordination.
- Improve the foundation for supporting or establishing these programs by:
    - Undertaking a coordinated effort among state and federal partners to assess critical risks, including anticipated sea-level rise, and increased frequency of storms leading to coastal inundation, to ensure that impacts are addressed at both local and regional scales.
    - Communicating available risk avoidance actions and providing resources about approaches to planning for individuals and communities.
    - Evaluating areas on a system-wide basis to determine the appropriateness of structural or non-structural risk reduction opportunities to include environmentally preferred alternatives.
- Explore opportunities for broadened implementation support for coastal improvement programs.

- Develop alternative implementation processes for coastal improvement projects that can be responsive to community needs.
- Evaluate opportunities to manage environmental mitigation on a watershed basis, with a goal of achieving the greatest overall environmental benefits.

Provide analytical support tools to enhance community planning, risk assessment and smart growth implementation

Many activities are underway in the Gulf to support communities as they examine current and future needs to enhance community resiliency. Coastal communities face significant risk of inundation from storms and sea-level rise. They also face a key challenge in their efforts to reduce this risk: accessing accurate and understandable information and decision-support tools about potential solutions. Models, forecasts and visualization tools will improve understanding of the impacts of coastal hazards and climate impacts on livelihoods and ecosystem services. Communities also need tools such as BMPs for local institutions about storm protection, fire protection, coastal building code improvements, stormwater runoff and nonpoint source pollution. In addition, evaluating coastal areas on a system-wide basis will help communities determine the appropriate portfolio of structural and non-structural risk reduction measures to include ecosystem restoration.

The concept of "smart growth" offers a viable framework for the local response planning efforts of Gulf Coast communities. Smart growth recognizes that communities should be economically vibrant and sustainable while providing a high quality of life and supporting ecosystem integrity. Communities that choose to implement coastal smart growth elements can do so through local zoning policies and building codes, local and regional planning activities, and incentives for private developers created with full stakeholder engagement.

Building on existing efforts, recommended actions include:

- Develop a toolbox of storm buffer options that could be considered by local entities.
- Inventory and evaluate models for storm surge, waves and coastal erosion to determine most appropriate/best ones for use in developing risk assessment of storm surge wave impact and sea-level rise.
- Through the Gulf of Mexico Alliance and the NOAA Coastal Storms Program, refine the risk or vulnerability indices (such as the Mississippi-Alabama Sea Grant Consortium's and the Louisiana Sea Grant's Community Resilience Index) to enhance the local understanding of risk from surge and waves, and to prioritize at-risk populations.
- Integrate and align available analytical tools and resources supporting Smart Growth application as well as potential opportunities for technical assistance. For example, EPA's Sustainable Communities Building Blocks program, the NOAA-EPA Coastal Community Development Partnership, Louisiana's Coastal Land Use Toolkit, and the HUD-DOT-EPA Partnership for Sustainable Communities help communities implement development approaches that protect the environment, improve public health, create jobs, expand economic opportunity and improve overall quality of life.

Enhance environmental education and outreach

Given the ever-changing nature of the Gulf ecosystem and the challenges that communities along the Gulf face, there is a need to increase public awareness of hazards, impacts to ecosystem services, and the interconnected nature of the ecosystem and the communities it supports. Programs that provide citizens with hands-on learning experiences increase understanding of links between the environment and their well-being. This understanding will enhance appreciation for the natural systems on which they depend and better prepare individuals and communities to face complex decisions about their interactions with their environment. This knowledge and awareness are essential for achieving environmental stewardship, which will help to ensure sustainable Gulf communities.

REGIONAL CLIMATE SERVICES PARTNERSHIP

NOAA is committed to promoting healthy ecosystems, communities and economies that are resilient in the face of changing climate conditions. NOAA's Regional Climate Services Partnership, which is made up of federal and non-federal partners such as Regional Integrated Science and Assessments, Regional Climate Centers, the National Integrated Drought Information System and Regional Climate Services Directors (RCSDs), seeks to assess regional needs and vulnerabilities, and then develop and deliver timely climate services that aid mitigation and adaptation choices. The Partnership also works to build a climate-literate public that understands its vulnerabilities so that it can appropriately plan ahead.

In fiscal year 2012, the RCSD for the Southern Region, working with NOAA's Gulf Coast Landscape Conservation Liaison, will lead an effort to develop a Gulf of Mexico climate services "roadmap" that will enhance NOAA-DOI interagency collaboration on climate services in this region. NOAA and DOI programs that focus on climate services, climate science and conservation delivery will be engaged in this effort. The focus of the "roadmap" effort in the Gulf will be a regional workshop that will provide an opportunity to highlight existing capacity and expertise, identify shared priority regional needs, and develop an integrated and coordinated approach to implementation of regional climate services activities.

Recommended actions include:

- Increase citizens' awareness of the Gulf by promoting engagement opportunities in their communities, including expanding opportunities for community-based restoration.
- Expand the use of public-private partnerships to enhance the development and delivery of Gulf of Mexico environmental educational programs.
- Include the economic value of the Gulf ecosystems in environmental education.
- Support and increase informal environmental education programs for Gulf of Mexico coastal communities, including reaching K-12 audiences.

- Leverage existing partnerships among state and federal agencies, affected Indian tribal governments, universities, nonprofits and community organizations to engage citizens in addressing current and future restoration and conservation needs, with particular attention to under-represented individuals and communities.

IV. Science-Based Adaptive Management

The Executive Order creating the task force highlights the critical need to ensure that restoration efforts have a robust scienti.c foundation. The need for science to support restoration and conservation is readily apparent, at a whole ecosystem level as well as at the project-speci.c level, as highlighted throughout many of the actions previously described in the Strategy. However, the dire state of many elements of the Gulf ecosystem cannot wait for scienti.c certainty and demand immediate action. A process is needed that allows for restoration efforts articulated here to move ahead in a scienti.cally defensible manner, increasing the fundamental scienti.c certainty necessary for successful restoration and expanding current knowledge of the state of the system. This process, and its associated resources, should be integrated into restoration planning and projects to ensure that the science is appropriately considered and suf.ciently supported. The process should determine the ef.cacy of the restoration actions through a focused effort of monitoring, modeling and research to support effective management and decision-making. More speci.c, preliminary monitoring, modeling and research needs are discussed in Appendix C.

> The dire state of many elements of the Gulf ecosystem cannot wait for scientific certainty and demand immediate action.

Adaptive management is a process of learning by doing, wherein flexibility is built into projects, and actions can be changed based on their progress toward a de.ned end-state. A key component is a feedback mechanism that is used to sequentially improve management actions so that management decisions are routinely adjusted to achieve program goals and objectives. Incorporating the best science and technology into project design and implementation would allow for restoration of the ecosystem and protection of coastal communities in light of future uncertainties (e.g., climate change). Critical elements include:

- *Establishing an effective adaptive management framework* with critical research, modeling and monitoring elements to support adaptive management. This framework can provide a foundation for decision-making; ensure that restoration efforts have clearly defined goals, objectives, and measures of success; provide long-term, continuous scientific data, analysis, and recommendations critical to the design, implementation, and monitoring of restoration and conservation projects; develop enabling tools, methodologies and protocols; resolve uncertainties that limit restoration planning; and assess the immediate and long-term effectiveness of restoration and conservation actions. This framework should ensure that restoration projects include assessment and integration of existing scientific knowledge and data, as well as an evaluation of past successes and failures of similar efforts. This framework would also provide for improved coordination and collaboration among federal and state agencies, industry (e.g., oil and gas, transportation), NGOs and stakeholders.

  Central to this framework is the establishment of an integrated, interdisciplinary, interagency effort, under the direction of the task force. This can include assembling necessary science working and advisory groups and teams, capitalizing on existing science consortiums, programs, universities, and institutions within Gulf states and across the region, to share resources and information, exchange ideas, identify concerns, and create solutions in the context of adaptive management for sustainability of Gulf Coast ecosystems. Additionally, implementation of adaptive management should be continuously improved by establishing a review and evaluation process that incorporates a critical assessment of the program and its effectiveness in supporting restoration and protection efforts. Independent scientific guidance and review will help ensure that robust science and peer review are fully integrated into restoration planning and the adaptive management process and that these efforts are routinely assessed and evaluated.
- *Establishing monitoring, modeling, and scientific research* to meet the scientific needs of the Strategy. Monitoring, modeling and research development activities should be integrated from the initial stages of restoration planning through to adaptive management decision-making.
- *Developing a comprehensive "watershed to Gulf" monitoring program.* Feedback from continuous, long-term, accurate monitoring provides the "adaptive" feature that is the basis of adaptive management and can be used to judge project effectiveness and the impact of collective restoration and conservation efforts across the Gulf ecosystem more broadly. This system should integrate and enhance existing comprehensive watershed, estuarine, coastal, and offshore monitoring networks and observing systems, including existing platforms and structures to improve cost-effective mapping, monitoring, and assessment of inland, estuarine, coastal and offshore environments. This program should also include identification and development of mechanisms for integrated management, synthesis and public availability of data and information, including use of and access to existing data, and development of data products and information services.
- *Establishing a Gulf of Mexico modeling network* to provide tools to increase certainty in forecasts and estimates of ecosystem function and services for decision-makers and the public. Models can be used to modify or adjust restoration and protection

actions, and to inform analysis and guidelines on the efficacy of different restoration strategies/projects (e.g., re-establishment or modification of freshwater flow, nutrient loads, suspended sediment deposition, storm buffers, barrier island restorations).

- *Investing in research and basic exploration* to understand the ecosystems in the Gulf and how they can be resilient to impacts from episodic events, such as hurricanes or oil spills, and long-term changes such as climate impacts. Humans are part of the ecosystem; focused research linking ecological and community resilience, including human impacts, solutions and risk, is needed. Research is particularly important in supporting management when it increases understanding of how physical, chemical and biological components interact with each other, helping scientists and managers evaluate known and potential impacts of restoration actions on environmental components and processes. Research should be directed at reducing scientific uncertainty to improve confidence in modeling and monitoring tools, and ultimately improve management actions. Identifying needs and implementing research should be included as part of restoration planning; research should be integrated into the adaptive management decision-making process, along with monitoring data and modeling outputs.
- *Providing decision support.* A central component of a strong adaptive management strategy is articulating and conveying results in a compelling and clearly understood format that can help inform effective decision-making and improve public understanding. Effective decision-support must also include a clear articulation of uncertainties and assumptions to help inform planning, assessment and trade-off analysis.
- *Developing integrated decision-support tools and systems,* including expansion and enhancement of predictive, simulation, and risk assessment models and ecological forecasting capabilities. (See "Integrated Ecosystem Assessments" and "Landscape Conservation Cooperatives.")
- *Developing decision-making visualization and data aids* that overlay the myriad uses of the Gulf that can potentially interact with energy and mineral development, such as Virtual Louisiana, Virtual Alabama and Gulf Data Atlas.
- *Expanding ecosystem services and benefits analysis tools and capabilities* to determine the socioeconomic benefits that ecosystems provide throughout the Gulf region.
- *Establishing indicators of success* and monitoring assessments to evaluate how well program elements meet their stated goals. The performance measures must be quantifiable and understandable to the public, have outcomes or targets specified for the desired Gulf condition, be sensitive to ecosystem change as a result of the Strategy implementation, and verify restoration and protection effectiveness and answer hypotheses.
- *Developing a Gulf-wide progress report.* This report should provide summary status information on ecosystem endpoints and communicate progress of management in improving ecosystem function (including ecosystem services). It should reflect trends over time to judge progress in an easy-to-understand format for the public and decision-makers, providing information on both how the ecosystem is functioning and why.

INTEGRATED ECOSYSTEM ASSESSMENTS

NOAA is adopting a comprehensive ecosystem-based management approach where the whole ecosystem, including its human dimensions, is considered at the same time. NOAA's approach to integrated ecosystem assessments (IEAs) is at ecological scales relevant to management questions, allowing resource managers to make more informed and effective decisions to achieve ecological and socioeconomic objectives. IEAs are a synthesis and quantitative analysis of information on relevant physical, chemical, ecological, and human processes in relation to specified management objectives.[119,120]

The national IEA program, which will include eight regions based on the U.S. Large Marine Ecosystems, has been initiated in three regions, including the Gulf of Mexico. The Gulf IEA will identify and accumulate relevant ecosystem data sets, and make them available through a data system and services framework for use in food web and ecosystem models. Trophic linkages data will be used in existing and future food web and ecosystem models. Ecosystem indicators will be developed and inform management through an Ecosystem Status Report.

LANDSCAPE CONSERVATION COOPERATIVES

In 2010, DOI established LCCs, forming a national network of public-private partnerships that provide shared science to ensure the sustainability of America's land, water, wildlife and cultural resources. LCCs provide scientific and technical support for landscape–scale conservation in an adaptive management framework by supporting conservation planning, providing decision support tools, prioritizing and coordinating research, and designing inventory and monitoring programs. Partner agencies, states and organizations coordinate with each other while working within their existing authorities and jurisdictions. In the Gulf of Mexico region, four LCCs are being established: *Gulf Coast Prairie*, which includes coastal Texas and Louisiana; *Gulf Coastal Plains and Ozarks*, which connects with the Louisiana coast and stretches across Mississippi, Alabama and the panhandle of Florida; *South Atlantic*, which includes the big bend area of Florida; and *Peninsular Florida*, which covers the rest of the Florida Gulf coast down through the Keys.

V. NEXT STEPS

The task force's primary purpose is to support and coordinate efforts of Gulf Coast states, the federal government, tribes and local governments to improve efficiency and effectiveness in the implementation of Gulf Coast ecosystem restoration actions. This Strategy is an initial, significant step in this effort. The Strategy sets forth a shared set of overarching goals based on existing goals, scientific assessments and stakeholder input, and proposes major actions for immediate and longer-term implementation.

In addition to mandating the task force's overall coordination role, Executive Order 13554 defines the task force's responsibilities to:

- Support the NRDA process by referring potential ecosystem restoration actions to the Trustee Council for consideration and facilitating coordination among the relevant departments and agencies as appropriate.
- Engage local stakeholders, communities and the public to ensure they have an opportunity to share their needs and viewpoints to inform the work of the task force.
- Identify major policy areas where coordinated intergovernmental action is necessary.
- Prepare a biennial update to the President on progress toward the goals of Gulf Coast ecosystem restoration as outlined in the Strategy.
- Provide leadership and coordination of research needs in support of ecosystem restoration planning and decision-making and facilitate consideration of relevant scientific and technical knowledge.
- Communicate with affected tribes in a manner consistent with Executive Order 13175.
- Coordinate with relevant agencies and offices on ways to encourage health and economic benefits associated with ecosystem restoration.

Going forward, the task force can serve as a valuable intergovernmental forum for senior policy officials at the state and federal levels to support efforts to restore the Gulf Coast ecosystem and address barriers to implementation such as science needs, regulatory complexities and resources. In this time of fiscal constraints, it is important for federal and state agencies to closely collaborate and align resources and actions to support projects that are high priority and ready to begin in the near future. The task force intends to work with other entities focused on Gulf Coast restoration to align and coordinate efforts; this includes the Gulf of Mexico Alliance, the National Ocean Council, research institutions and others. Continued work is needed to translate some of these recommendations into action on the ground and identify ways to track restoration progress. The task force plans to build on work already done by developing an implementation framework with specific milestones.

Several of the actions described in this Strategy will be immediately implemented by one or more members of the task force. Other actions address long-standing challenges in the Gulf that will require more focused effort over time to determine the course of action. Following the final release of the Strategy, the task force intends to begin immediately to develop the activities, outcomes, performance measures, milestones and short-, medium- and long-term tasks necessary to implement this Strategy. The task force intends to complete work on defining outcomes and performance measures for the actions described in this Strategy within six months after the final Strategy is released.

Task force members need to work to more effectively align programs, resources and science to support restoration planning, design, and implementation and will develop tools to measure progress in achieving ecosystem restoration goals. By ensuring greater coordination at all levels of government, the task force can promote more effective and efficient management of restoration efforts and support project proponents.

To this end, the task force intends to collaborate with member states and agencies to implement the actions set forth in this Strategy. It also intends to establish mechanisms in the near term to facilitate more efficient development and implementation of restoration projects. These efforts should enhance coordination, prioritize ecosystem restoration actions, facilitate leveraging of funds and improve restoration permitting and regulatory review of the programs supporting the Strategy. Many of the actions outlined in the Strategy are currently supported by ongoing plans and supporting programs. The initial focus will be to identify and build upon these efforts. For actions not currently supported, the task force intends to:

- Specify the current status of each issue, including which agency is working on the issue, under which authority, and with which resources.
- Identify a lead agency, state or other actor for each task needed to implement the action, along with appropriate partners.
- Create a schedule of specific activities to accomplish each goal, along with any necessary or available resources, authority, science or external support needed.
- Establish a science-based approach for ensuring success of actions undertaken. Key elements include:
  - Building a robust interdisciplinary, interagency effort to guide implementation of project-specific and ecosystem-wide restoration adaptive management
  - Establishing critical monitoring, modeling and research elements that provide the scientific foundation for the restoration goals outlined in the Strategy.
  - Developing analyses and tools that support well-informed, timely decisions based on an up-to-date understanding of ecosystem trends over time, future conditions, effectiveness of past and ongoing restoration efforts, and necessary changes or considerations for future restoration efforts.
  - Developing a science advisory capacity to ensure that peer review, science and scientifically based findings are properly incorporated into the adaptive management and decision-making process.
- Provide critical decision-support tools to reflect ecosystem trends over time, communicate progress of restoration projects, and inform critical decisions and necessary changes in restoration efforts.
- Continue to engage the public and communities on the restoration effort by providing a forum for the public to bring its ideas and needs regarding ecosystem restoration to governmental leaders. A citizens' advisory committee to advise the Chair of the task force has been approved.
- Establish and maintain the core interagency staff expertise needed to develop the implementation plan and guide its execution as envisioned in this Strategy.
- Expand opportunities for collaboration with local governments, business entities, and NGOs as a critical component of Strategy implementation.

The task force envisions that the programs necessary to implement the Strategy will be implemented by the appropriate state or federal agency or agencies. Some programs identified as necessary to achieve restoration might require specific state or Congressional authorization.

The task force can undertake critical work, as outlined above, to ensure that the tools and relationships needed for successful Gulf Coast restoration are in place or under development. By assisting other ongoing restoration programs and bringing a comprehensive ecosystem focus to Gulf Coast restoration efforts, underpinned by good science and effective stakeholder involvement, the task force strives to fulfill a very important mission: to put America's Gulf Coast on the road to long-term recovery and restoration.

The task force reaffirms the importance of Secretary Mabus' two initial recommendations: that Congress (1) set aside a significant portion of any potential Clean Water Act civil penalties associated with the *Deepwater Horizon* oil spill for recovery in the Gulf and (2) create a Gulf Coast Recovery Council to coordinate restoration and recovery in the region as an eventual successor to this task force.

The Executive Order directed that this Strategy describe the circumstances under which termination of this task force would be appropriate. The task force believes that in the event Congress passes legislation establishing a Gulf Coast Recovery Council or similar body with comparable responsibilities, the task force would terminate and facilitate any needed transition. There could be other conditions under which the task force is no longer needed and it may, by consensus of its members, choose to forward to the President a recommendation to cease its operations. However, there is currently a clear need for federal/state cooperation in support of the actions described in this Strategy. The task force intends to revisit this question before the close of fiscal year 2012.

APPENDIX A. EXECUTIVE ORDER 13554

Presidential Documents

Executive Order 13554 of October 5, 2010

Establishing the Gulf Coast Ecosystem Restoration Task Force

By the authority vested in me as President by the Constitution and the laws of the United States of America, it is hereby ordered as follows:

Sec. 1. *Purpose.* The Gulf Coast is a national treasure. Its natural resources are an important economic engine for the entire United States; its waters sustain a diverse and vibrant ecosystem; and the Gulf's culture, natural beauty, and historic significance are unique. Each year, millions of tourists visit the Gulf to vacation, swim, boat, fish, hunt, and bird-watch; and, together, the Gulf's tourism and commercial and recreational fishing indus-tries make a significant contribution to the United States economy. More than 90 percent of the Nation's offshore oil and gas is produced in the Gulf, and it is where nearly one-third of seafood production in the continental United States is harvested.

The United States needs a vibrant Gulf Coast, and the Federal Government is committed to helping Gulf Coast residents conserve and restore resilient and healthy ecosystems in the Gulf of Mexico and surrounding regions that support the diverse economies, communities, and cultures of the region. To effectively address the damage caused by the BP Deepwater Horizon Oil Spill, address the longstanding ecological decline, and begin moving toward a more resilient Gulf Coast ecosystem, ecosystem restoration is need-ed. Ecosystem restoration will support economic vitality, enhance human health and safety, protect infrastructure, enable communities to better with-stand impact from storms and climate change, sustain safe seafood and clean water, provide recreational and cultural opportunities, protect and preserve sites that are of historical and cultural significance, and contribute to the overall resilience of our coastal communities and Nation.

In order to achieve these objectives, it is necessary that Federal efforts be efficiently integrated with those of local stakeholders and that particular focus be given to innovative solutions and complex, large-scale restoration projects. Efforts must be science-based and well-coordinated to minimize duplication and ensure effective delivery of services. This order establishes a Gulf Coast Ecosystem Restoration Task Force to coordinate intergovern-mental responsibilities, planning, and exchange of information so as to better implement Gulf Coast ecosystem restoration and to facilitate appropriate accountability and support throughout the restoration process.

Sec. 2. Establishment of the Gulf Coast Ecosystem Restoration Task Force. There is established the Gulf Coast Ecosystem Restoration Task Force (task force).

(a) The task force shall consist of:
  (1) A senior official from each of the following executive departments, agencies, and offices, selected by the head of the respective department, agency, or office:
    a. the Department of Defense;
    b. the Department of Justice;
    c. the Department of the Interior;
    d. the Department of Agriculture;
    e. the Department of Commerce;
    f. the Department of Transportation;
    g. the Environmental Protection Agency;
    h. the Office of Management and Budget;
    i. the Council on Environmental Quality;
    j. the Office of Science and Technology Policy;
    k. the Domestic Policy Council; and
    l. other executive departments, agencies, and offices as the President may, from time to time, designate.
  (2) Five State representatives, appointed by the President upon rec-ommendation of the Governors of each Gulf State, who shall be elected officers of State governments (or their designated employees with authority to act on their behalf) acting in their official capacities.
(b) The task force may include representatives from affected tribes, who shall be elected officers of those tribes (or their designated employees with authority to act on their behalf) acting in their official capacities. The task force shall, in collaboration with affected tribes, determine an appropriate structure for tribal participation in matters within the scope of the task force's responsibilities.
(c) The President shall designate a Chair of the Task Force from among senior officials of executive departments, agencies, and offices represented on the task force. The Chair shall lead the coordination of intergovernmental Gulf Coast ecosystem restoration efforts and oversee the work of the task force. The Chair shall regularly convene and preside at meetings of the task force, determine its agenda, and direct its work. The Chair's duties shall also include:

(1) facilitating a smooth transition from the response phase of addressing the BP Deepwater Horizon Oil Spill to the restoration phase;
(2) communicating and engaging with States, tribes, local governments, other stakeholders in the Gulf Coast region, and the public on ecosystem restoration, as well as other aspects of Gulf recovery, including economic recovery and public health efforts; and
(3) coordinating the efforts of executive departments, agencies, and offices related to the functions of the task force.

(d) Representatives of the Gulf States under subsection (a)(2) of this section shall select from among themselves a Vice-Chair of the Task Force.

Sec. 3. *Functions of the Task Force.* The task force shall be an advisory body to:

(a) coordinate intergovernmental efforts to improve efficiency and effectiveness in the implementation of Gulf Coast ecosystem restoration actions;
(b) support the Natural Resource Damage Assessment process by referring potential ecosystem restoration actions to the Natural Resource Damage As-sessment Trustee Council for consideration and facilitating coordination among the relevant departments, agencies, and offices, as appropriate, subject to the independent statutory responsibilities of the trustees;
(c) present to the President a Gulf of Mexico Regional Ecosystem Restora-tion Strategy (Strategy) as provided in section 4 of this order;
(d) engage local stakeholders, communities, the public, and other officials throughout the Gulf Coast region to ensure that they have an opportunity to share their needs and viewpoints to inform the work of the task force, including the development of the Strategy;
(e) provide leadership and coordination of research needs in support of ecosystem restoration planning and decisionmaking in the Gulf Coast region, and work with existing Federal and State advisory committees, as appro-priate, to facilitate consideration of relevant scientific and technical knowl-edge;
(f) prepare a biennial update for the President on progress toward the goals of Gulf Coast ecosystem restoration, as outlined in the Strategy;
(g) communicate with affected tribes in a manner consistent with Executive Order 13175 of November 6, 2000, on consultation and coordination with Indian tribal governments; and
(h) coordinate with relevant executive departments, agencies, and offices on ways to encourage health and economic benefits associated with proposed ecosystem restoration actions.

Sec. 4. *Gulf of Mexico Regional Ecosystem Restoration Strategy.*

(a) Within 1 year of the date of this order, the task force shall prepare a Strategy that proposes a Gulf Coast ecosystem restoration agenda, including goals for ecosystem restoration, development of a set of performance indicators to track progress, and means of coordinating intergovernmental restoration efforts guided by shared priorities. In developing the Strategy, the task force shall:

(1) define ecosystem restoration goals and describe milestones for making progress toward attainment of those goals;
(2) consider existing research and ecosystem restoration planning efforts in the region, including initiatives undertaken by the National Ocean Council and the Mississippi River/Gulf of Mexico Watershed Nutrient Task Force (Gulf Hypoxia Task Force), in order to identify planning and restoration needs and ways under existing authorities to address those needs;
(3) identify major policy areas where coordinated intergovernmental action is necessary;
(4) propose new programs or actions to implement elements of the Strategy where existing authorities are not sufficient;
(5) identify monitoring, research, and scientific assessments needed to support decisionmaking for ecosystem restoration efforts and evaluate existing monitoring programs and gaps in current data collection; and
(6) describe the circumstances under which termination of the task force would be appropriate.

(b) The executive departments, agencies, and offices enumerated in section 2(a)(1) of this order shall, to the extent permitted by law, consider ways to align their relevant programs and authorities with the Strategy.

Sec. 5. *Administration.*

(a) The task force shall have a staff, headed by an Executive Director, which shall provide support for the functions of the task force.
(b) The Executive Director shall be selected by the Chair and shall super-vise, direct, and be accountable for the administration and operation of the task force.
(c) The Departments of Commerce (through the National Oceanic and Atmospheric Administration), the Interior (through the Fish and Wildlife Service), and Justice shall identify linkages and opportunities for the task force to complement the restoration progress of the Natural Resource Damage Assessment Trustee Council.
(d) At the request of the Chair, executive departments and agencies, includ-ing the Departments of Labor, Health and Human Services, Energy, and Homeland Security, the Small Business Administration, and the National Science Foundation, shall serve in an advisory role to the task force on issues within their expertise.
(e) The task force may establish such technical working groups as nec-essary to support its function. These working groups may include additional representatives from State and tribal governments, as appropriate, to provide for greater collaboration.
(f) The first meeting of the task force shall be held within 90 days of the date of this order.

Sec. 6. *Definitions.*

(a) ''Affected tribe'' means any Indian tribe, band, nation, pueblo, village, or community that the Secretary of the Interior acknowledges to exist as an Indian tribe as defined in the Federally Recognized Tribe List Act of 1994 (25 U.S.C. 479a(2)), physically located in a Gulf State.

(b) ''Ecosystem restoration'' means all activities, projects, methods, and procedures appropriate to enhance the health and resilience of the Gulf Coast ecosystem, as measured in terms of the physical, biological, or chemical properties of the ecosystem, or the services it provides, and to strengthen its ability to support the diverse economies, communities, and cultures of the region. It includes activity that initiates or accelerates the recovery of an ecosystem with respect to its health, integrity, and sustainability. It also includes protecting and conserving ecosystems so they can continue to reduce impacts from tropical storms and other disasters, support robust economies, and assist in mitigating and adapting to the impacts of climate change.

(c) ''Gulf State'' means any of the States of Texas, Louisiana, Mississippi, Alabama, and Florida.

(d) ''Natural Resource Damage Assessment'' means the process of collecting and analyzing information to evaluate the nature and extent of natural resource injuries resulting from the BP Deepwater Horizon Oil Spill and to determine the restoration actions needed to bring injured natural resources and services back to baseline conditions and make the environment and public whole for interim losses as defined in 15 CFR 990.30.

(e) ''Natural Resource Damage Assessment Trustee Council'' means the designated Federal, State, local, and tribal trustees as provided in 33 U.S.C. 2706, with trusteeship over natural resources injured, lost, or destroyed as a result of the BP Deepwater Horizon Oil Spill.

Sec. 7. *General Provisions.*

(a) To the extent permitted by law and subject to the availability of appropriations, the department, agency, or office rep-resented by the Chair shall provide the task force with such administrative services, funds, facilities, staff, and other support services as may be nec-essary for the task force to carry out its function.

(b) In addition to staff provided by the department, agency, or office represented by the Chair, other executive departments, agencies, and offices represented on the task force are requested to make services, staff, and facilities available to the task force for the performance of its function to the maximum extent practicable, to the extent permitted by law and subject to the availability of appropriations.

(c) Members of the task force shall serve without any additional com-pensation for their work on the task force.

(d) Nothing in this order shall be construed to impair or otherwise affect: (i) authority granted by law to an executive department, agency, or the head thereof, or the status of that department or agency within the Federal Government; or (ii) functions of the Director of the Office of Management and Budget relating to budgetary, administrative, or legislative proposals.

(e) Nothing in this order shall interfere with the statutory responsibilities and authority of the Natural Resource Damage Assessment Trustee Council or the individual trustees to carry out their statutory responsibilities to assess natural resource damages and implement restoration actions under 33 U.S.C. 2706 and other applicable law.

(f) This order shall be implemented consistent with applicable law and subject to the availability of appropriations.

(g) This order is not intended to, and does not, create any right or benefit, substantive or procedural, enforceable at law or in equity by any party against the United States, its departments, agencies, or entities, its officers, employees, or agents, or any other person.

The White House,
October 5, 2010.

APPENDIX B. THE GULF COAST STATES

Introduction

The restoration strategy outlined in this document employs a comprehensive ecosystem-based, regional approach. At the same time, the Strategy builds on activities already under way within each of the five Gulf Coast states, as well as collaborative activities among the states. The Strategy acknowledges the fact that while the states have a shared coastline and ecosystem, they also have unique and distinct natural features, cultural heritages, and economic resources. While effective ecosystem restoration will transcend state boundaries, the regional strategy should also be based on an understanding of—and be supportive of—state-specific efforts and perspectives.

Representatives of the five Gulf Coast states were fully involved in the development of the overall Strategy presented in this document. They also have developed the information presented in this appendix. The following sections identify the ecological and economic resources of each state, along with the priority actions being undertaken in each state (organized in terms of the Strategy's four goal areas). As the task force develops the supporting implementation plan in the coming months, it will work to find ways, where appropriate, to incorporate, support, and/or harmonize the state-based activities described herein within the context of the overall Strategy.

Alabama

Background

Alabama's coast is located along the northern Gulf of Mexico in Mobile and Baldwin Counties. The state's 607 miles of shoreline include the beach and dune systems along the Gulf of Mexico, approximately 400,000 acres of bay and estuarine waters,[121] and approximately 127,000[122] acres of various types of wetlands: fresh and salt marshes, scrub-shrub, forested and grass beds.

The current population of the state is 4,779,736. The population of Mobile County is 412,992, and the population of Baldwin County is 182,265. Overall, the state population increased 7.5 percent since 2000, while Mobile County increased 3.3 percent, and Baldwin

County increased 29.8 percent over the same timeframe.[123] Mobile and Baldwin Counties sit at the crossroads for expanded economic development for the northern Gulf of Mexico. They also serve as a major transition zone, where freshwater rivers mix with the tidally influenced salt water of the Gulf of Mexico.

Alabama's coastal area contains four broad natural ecosystems—terrestrial, freshwater, estuarine and marine/continental shelf—which support an extremely diverse assemblage of plants and animals. In terms of biodiversity (the number of distinct species in a given area), Alabama ranks fifth among states in the United States and first among those east of the Mississippi River.[124,125]

Within these ecosystems, Alabama has a profuse diversity of natural habitats, including rich sediments, seagrass beds, barrier islands, freshwater and saltwater wetlands, pitcher plant bogs, bottomland hardwood forests, wet pine savannas and upland long leaf pine and oak forests. The wealth of these habitats is what makes the Alabama coast unique and draws residents and visitors alike.

Alabama's coast is valued for its protective functions, natural resources and the uses and activities that take place on or near the water's edge, all of which contribute to the economy of the state and the entire nation.

The natural protective functions of Alabama's beach and dune systems, estuaries and wetlands are instrumental in preserving and enhancing ecosystems and in reducing risks to infrastructure and commercial and residential developments. The coastal areas also protect inland regions from erosion, flooding and storms.

Gulf and Meaher State Parks

Two state parks are located along the Alabama shoreline:

- *Gulf State Park* is a 6,000-acre park with a 5-mile stretch of sandy beach on the Gulf of Mexico. The park grounds contain small lakes, a large area of wetlands and a maritime forest. There are backcountry nature trails, a 1,520-foot fishing pier, cabins and campsites. The park has 2.5 million visitors and a quarter million overnight guests per year.
- *Meaher State Park* is a 1,327-acre park situated in the wetlands of Mobile Bay. It is a day-use picnicking and scenic park and has camping and bathhouse facilities for overnight guests. There is a boat ramp, fishing pier, and two nature trails that include a boardwalk with a close view of the Mobile Delta.

The area's natural resources include the Mobile-Tensaw Delta, which is an estimated 280-square-mile area with approximately 20,000 acres of open water, 10,000 acres of fresh-mixed marsh, 69,000 acres of swamp, and 85,000 acres of mixed bottomland forest. It is considered a principal state asset.[126] The Delta, along with Alabama's beaches and dunes, estuaries and wetlands, serves as a breeding and nursery ground for numerous species of finfish, shellfish, waterfowl, migratory birds and other wildlife, including many endangered and threatened species; thus, the coastal area is important in their life cycles. The estuaries of coastal Alabama are economically and environmentally important because of their exceptional biological diversity and productivity.

The Mobile Bay Watershed is the sixth largest river basin in the United States and the fourth largest in terms of stream flow.[127] It drains water from three-quarters of Alabama and portions of Georgia, Tennessee and Mississippi into Mobile Bay. Mobile Bay is also a point of entry for hundreds of smaller recreational and commercial vessels that traverse the lower Mobile Delta. Many of these vessels make the 450-mile trip to the Tennessee River through the inter-basin connector known as the Tennessee-Tombigbee waterway; other vessels cruise to other inland Alabama ports via extensive navigation projects on the Alabama and Warrior River systems.

Seventy-five percent of America's commercial fish catch and 80 to 90 percent of the recreational fish catch are dependent on estuaries, such as those found in Alabama, including Mobile Bay, Mississippi Sound, Weeks Bay and Perdido Bay.[128] The value of Alabama's commercial fishery and recreational boating and fishing statistics are recorded by the Alabama Department of Conservation and Natural Resources. In 2009, records showed 14,291 commercial trips landing 29.7 million pounds of seafood worth over $40.5 million. At present, 271,523 registered boats and 597,785 boat operator licenses are on file in the state. In 2009, 100,290 resident and non-resident saltwater fishing licenses were also issued.

Alabama's coastal resources also support a thriving tourism industry. The state reported that in 2009 more than 7.1 million visitors enjoyed the state's beaches, scenic vistas, access to the Gulf of Mexico and bays and waterways. In 2009, these visitors spent over $3.2 billion in Mobile and Baldwin Counties. These two counties rank first (Baldwin at 25 percent) and third (Mobile at 9.7 percent) in total travel-related employment for the state. In 2009, the total travel-related employment for the two counties was 56,294 employees, and total travel-related earnings by employees were over $1.2 billion.[129]

Essential to Alabama's coastal economy is the Port of Mobile, which is the global deepwater gateway for the state. In 2009, the port ranked 12th in the nation in the amount of total tonnage shipped.[130] The most frequent import and export commodities transferring through the port are coal, aluminum, iron, steel, lumber, wood pulp and chemicals. The Alabama State Port Authority's Mobile Container Terminal (APM Terminals Mobile) provides containerized cargo shippers with access to global networks covering all possible trade routes to and from the Port of Mobile. The port is served by 12 shipping lines offering fixed daily, weekly, bimonthly and monthly ocean carriage services to global destinations. APM Terminals Mobile operates the container terminal, which currently has five fixed-day, weekly vessel calls by Maersk, CMA CGM, Mediterranean Shipping, ZIM, APL and HMM lines. General cargo and project ocean carriage is provided by Grieg Star Shipping, Gearbulk, Westfal-Larsen Lines, C.I.C.-Clipper, Spliethoff, C.G. Railway/ International Shipholding, and MWC shipping. With a large number of ship berths with both 40-foot and 45-foot ship channel drafts and a new 1,175-by-715-foot turning basin,[131] the port can accommodate most ship sizes, including post-Panamax-sized vessels. The seaport's public terminals are located just three and one half hours from deepwater U.S. Gulf navigation.[132]

The port, combined with the city of Mobile's industrial center, represents a multi-billion-dollar impact on the regional, state and world economies. The port contributes 66,617 jobs and over $7.92 billion to Alabama's economy annually. In addition, virtually every conceivable service for the maritime industry can be found in the city of Mobile, including barge fleeting, stevedoring, towing, marine surveying, shipbuilding and ship repair, heavy lift and salvage, freight forwarding and Customs House brokerage firms, London Metals Exchange-certified warehousing, Foreign Trade Zones, vessel husbandry and chandlery, industrial diving, marine fumigation, maritime waste disposal, line handling, marine instrument calibration, dunnage removal, container and chassis repair and leasing, fuel bunkering, and many other services.[133]

Alabama is rich in energy resources and has considerable natural gas reserves. Offshore natural gas, shipbuilding and ship repair are among the coastal area's expanding businesses, creating a synergy for growth. As offshore oil and gas exploration continues, the area's shipbuilders are building offshore supply and rig-tending vessels and repairing rigs at their facilities on the Mobile River.[134] A number of local shipbuilders also build, renovate and repair vessels of all sizes for the U.S. military.

State Priorities

The Task Force Strategy lays out four overarching goals. Alabama's priority actions for each goal are described below.

Restore and Conserve Habitat

The natural habitats of Alabama's coast have experienced changes, primarily through loss, over the last 100 years. Some of these losses are a result of natural events; others are due to human activities, such as population growth, land-use conversion, shoreline hardening, introduction or transfer of invasive species, and runoff and excess sediments. These natural ecosystems are complex and heavily impacted by patterns of land use and changing land cover.

The beaches of Gulf Shores, Orange Beach, Gulf State Park and Dauphin Island have all experienced major erosion in the past decade,[135] starting with Tropical Storm Isidore in September 2002, and Hurricanes Ivan and Katrina in September 2004 and August 2005, respectively. In 2005 and 2006, Alabama placed nearly 8 million cubic yards of sand along more than 15 miles of beach—the largest beach nourishment project in the state's history. This project, and other beach nourishment efforts, have provided additional habitat, storm protection benefits and better recreational capacity. However, these projects totaled more than $31 million, and the cost of properly maintaining them will continue over the years.[136] Accordingly, identifying a consistent source of funding is crucial for their sustainability.

Many successful habitat restoration efforts have also been conducted in the state. Small-scale projects include the restoration of salt marsh and living shoreline projects at Helen Wood Park, numerous marsh planting projects conducted by the "Grasses in Classes Program" in conjunction with the Weeks Bay National Estuarine Research Reserve, the Mobile Bay Oyster Gardening Program, and a number of experimental oyster reef projects. Large-scale projects include the Little Bay Project and the Coffee Island/Alabama Port Shoreline Protection Project, which have restored over 30 acres of salt marsh and/or protected approximately 3 miles of shoreline from further loss from erosion, as well as providing important habitat for oysters, finfish and shellfish. The Coffee Island/Alabama Port Project is a prime example of federal and state partners working with academic, scientific and nonprofit institutions, such as the Dauphin Island Sea Lab, the Mobile Bay NEP, Mobile Baykeeper and The Nature Conservancy. The Alabama Department of Conservation and Natural Resources has also established an oyster reef that is expected to be self-sustaining under appropriate monitoring and management. However, the true success of such restoration projects depends upon dedicated funding for monitoring and ongoing, long-term maintenance needs.

Alabama's priorities for conserving and restoring habitat include:

- Work with the federal government to establish support for and development of a land protection strategy for the Alabama Gulf Coast.
- Develop a strategy and support programs for acquiring and preserving endangered coastal habitat by working with existing state and federal programs for land acquisition.
- Restore wetlands through the use of dredged sediments and identify physical and financial resources to restore critical habitats.
- Restore barrier islands and sandy beaches through beneficial use of dredged material and beach renourishment efforts and identify physical and financial resources to restore critical habitats.
- Enhance existing oyster reefs by planting cultch materials in areas affected by storms, and establish new oyster reefs by planting cultch and relocating live oysters from other areas.
- Implement policy changes to protect freshwater wetlands; value ecosystem services; restore streams, rivers, and creeks; and improve management of stormwater runoff.
- Protect and stabilize salt marsh edges, manage wave energy to deter erosive impacts and restore salt marshes in strategic locations.
- Implement policy changes to encourage living shoreline technologies, and demonstrate appropriate and affordable technologies to private landowners.
- Adopt and implement a statewide plan for the long-term management of invasive species.
- Document patterns of land use and land cover changes and how they will impact habitats in the foreseeable future.
- Enhance and create offshore and inshore natural and artificial fish habitat and structures.
- Review and assess cumulative impacts of dams and other infrastructure that may impact freshwater and sediment inputs.
- Coordinate with the federal government to establish habitat restoration projects that require no federal match; when federal match is required, coordinate with federal agencies to allow for creative match options.

Restore Water Quality

Maintaining adequate supplies of clean water is critical to the high quality of life enjoyed by the citizens of Alabama. Water is necessary for maintaining agricultural production, industrial processes, power generation and public health. The Alabama Gulf Coast also is a main source of recreation for both citizens and visitors. As Alabama grows, the demand for clean water will continue to increase.

Major water quality issues in the Mobile River Basin include nutrient enrichment, sedimentation, pesticides and toxins, habitat degradation, heavy metals, bacterial contamination and the health of the estuarine environment and its fisheries. To track changes in the condition of this basin, the State of Alabama has supported a variety of environmental monitoring programs, including water quality, habitat change and key living resource populations. This monitoring will establish long-term datasets to track change over time.

Alabama's priorities for restoring water quality include:

- Reduce excessive sediment and pathogen loads in waterways through improved management of stormwater runoff, restoration of inland waterways (including streams and creeks) and promotion of sustainable land use policies.
- Improve and enhance infrastructure that promotes centralized wastewater treatment systems to improve water quality.
- Improve understanding of impacts from heavy metals and mercury and how to cycle these metals out of sediments.
- Expand public education and incentives related to nonpoint source pollution and promote local management alternatives.
- Improve water quality and clarity to promote seagrass restoration.

Replenish and Protect Living Coastal and Marine Resources

To protect the rich diversity of coastal and marine resources in the state, Alabama is focusing on gaining a better understanding of the history, habitat requirements, life cycles, and strengths and weaknesses of native species. The state is also examining problems associated with the introduction of exotic species and working to ensure the continued health of commercial and recreational fisheries.

The Alabama coast is recognized for its recreational fishing opportunities and commercial fishing industry. Many of the species that support these fishing opportunities have complex life histories. They usually exhibit onshore/offshore migrations and a strong dependence on estuarine habitats during juvenile stages. Long-term monitoring of estuarine-dependent organisms provides insight into the status of fishery resources, which could help to determine the effectiveness of habitat restoration programs, the consequences of habitat degradation, and the impacts of invasive species.

The state's priorities for protecting and replenishing living coastal and marine resources include:

- Increase large-scale funding for oyster reef restoration and artificial reef creation and enhancement.
- Increase funding for marine resource research and fishery and shellfish production facilities.
- Improve temporal and spatial monitoring of living and coastal marine resources, including nursery habitats and dependent faunal species.
- Develop adaptive policies for the management of supply and demand of commercial and recreational fishery species.
- Maintain and enhance fishery stocks in nearshore, offshore and coastal Alabama waters.
- Promote, develop, monitor and enhance the artificial reef zones in both inshore and offshore Alabama waters.

Enhance Community Resilience

The coastal region from east of the Mississippi River to east of Mobile Bay is more likely to be affected by a moderate to severe climatic event than other area in the Gulf. Over a 15-year period, nine storms have caused some degree of damage in Alabama. These storms include seven named hurricanes—Erin, Opal, Danny, Georges, Ivan, Dennis and Katrina (six of which were Category 2 or higher), and two named tropical storms—Hanna and Isidore. Hurricane Ivan alone caused over $2 billion in insured losses in 2004.[137] While not every storm was a direct hit, they all caused problems. For example, Tropical Storm Isidore, which made landfall in coastal Louisiana, caused significant coastal beach erosion in Alabama—in fact, the shoreline prior to that storm has served as the restoration standard for local beach nourishment efforts.

The *Deepwater Horizon* oil spill of 2010 is just the latest of a series of disasters to strike the Alabama coast, and it is evident from these events that the entire coastal area of Alabama needs a comprehensive plan to reduce the significant risk and damages to public safety, property, economic viability and environmental resources from future natural and human-made stressors. Being a coastal state brings repeated and ongoing challenges—it goes with the territory. For Alabama, it is essential to understand the cumulative impacts of these stressors, which have long-lasting effects on the natural resources, physical resources and local economies.

A comprehensive plan would ensure the economic and social vitality of the state's coastal communities and their ability to survive crisis, influence change and be healthy places for their residents and visitors. Toward this end, the Coastal Recovery Commission of Alabama recently drafted a roadmap to guide the development and implementation of strategies across a broad range of categories to strengthen the region's communities and their adaptability and sustainability over time. A comprehensive plan for coastal Alabama could be patterned after the successful MsCIP, which acknowledges the importance of a healthy coastal ecosystem to the sustainability and resiliency of both the natural and human-made infrastructure of the coast. A comprehensive plan could also be crafted that would build upon the Coastal Recovery Commission's work in identifying the issues facing coastal Alabama and move toward possible solutions:

- Creating opportunities for the collaboration of local, state and federal agencies and Alabama coastal residents to maximize the use of resources in support of the comprehensive planning effort.
- Reducing the susceptibility of residential, commercial and public infrastructure to storm damages.
- Improving habitats for coastal and marine resources to support commercial and recreational harvest.
- Assisting in the recovery of natural and human-made features damaged by erosion or unwise land use and development decisions.
- Promoting long-term erosion reduction during future natural hazards.
- Promoting diversification of economies within the two coastal counties as a means of economic resilience from future hazards.

All of the above could be included in an Alabama Coastal Resiliency Plan. Currently, there are no existing authorities under which the U.S. Army Corps of Engineers could participate in such a comprehensive planning effort. However, the state has held meetings with the U.S. Army Corps of Engineers to explore possible partners and funding options. The most promising partner is the Alabama Department of Conservation and Natural Resources, whose State Lands Division currently houses a number of coastal programs; a MsCIP-like comprehensive planning effort would fit well within the Division's existing programs. Other likely sponsors include Mobile and Baldwin Counties. Other partners that would be included in the planning effort are the local municipalities, the Mobile Bay NEP, the Alabama State Port Authority, and NGOs such as Mobile Bay Keeper, Smart Coast Inc., Alabama Coastal Foundation, Partners for Environmental Progress, Envision Coastal Alabama, Mobile Area Chamber of Commerce, Eastern Shore Chamber of Commerce, South Baldwin Chamber of Commerce, Gulf Coast Area Chamber of Commerce, Gulf Shores/Orange Beach Tourism, and the Mississippi-Alabama Sea Grant Consortium.

The state's priorities for enhancing community resilience include:

- Promote the development of community conservation plans that include elements of coastal management, emergency response and community development.
- Improve knowledge of the dollar value associated with environmental protection in terms of long-term community sustainability and resilience.
- Enhance opportunities to promote the understanding of the necessity of economic resilience as a component of community resilience through various programs, such as those offered through Gulf State Park, which increase both visitation to and appreciation for the state's coastal community.
- Implement an Alabama Coastal Resiliency Plan in the aftermath of catastrophic events.
- Enhance and promote tourism and economic development in the coastal area.
- Promote, restore and enhance public access and recreation to coastal resources.

Florida

Background

Florida forms the southernmost terminus of the United States and the eastern boundary of the Gulf of Mexico; it has three miles of territorial seas off the Atlantic and three leagues (approximately 10 miles) in the Gulf. Along its Gulf Coast alone, Florida has 5,095 miles of tidal shoreline,[138] 436 miles of sandy beaches,[139] and over 7.4 million acres of tidally submerged lands.[140]

Florida's population numbers 18,801,310 people. Statewide, 80 percent of this population lives along the coast, and no point in the state is more than 60 miles from the coast. Twenty-three of Florida's 67 counties border the Gulf Coast, with a combined population of more than 6.6 million residents.[141]

Ecologically, the state is very diverse. Geography heavily influences this diversity of coastal and marine habitat, creating essentially an island state, almost entirely surrounded by sea. From temperate Pensacola to tropical Key West, the nearly 900-mile journey crosses seven degrees of latitude through rare coastal dune lakes, pine flatwoods and prairies, scrub, coastal hammocks, a variety of wetlands, and the Florida Everglades and Florida Keys' coral reefs.

Along the Gulf Coast, from Pensacola to Apalachicola, barrier islands help form tidal estuaries near river mouths where hardwood swamps transition to salt marsh wetlands. From Ocklockonee Bay to Crystal River, crystal clear springs and blackwater swamp forests feed streams and rivers that empty directly to the Gulf Coast, where vast and important seagrass beds cover the sea floor. From Tarpon Springs south, mangroves replace salt marsh and form fringing forests and inland tidal swamps.

The broad, shallow continental shelf along Florida's Gulf Coast supports extensive tropical coral reefs that stretch from the Florida Keys to the Snapper Banks off Pensacola. Strategically placed artificial reefs supplement these natural coral reefs. Together, these coral communities support an incredible diversity of fish and other marine life, including fish populations found nowhere else in the Gulf. The beauty and biodiversity of Florida's coral reefs attract divers and fishermen from around the world. The state, particularly the Florida Keys, is recognized as a diving capital of the world. The reefs, along with intertidal oyster bars, barrier islands, tidal salt marsh, mangroves and submerged seagrass meadows provide a buffer against storms and hurricanes, making adjacent coastal communities safer and more resilient.

The Gulf Loop current is significant for the biodiversity of state resources as it circulates the warm tropical seawater from the Caribbean that is partially responsible for establishing and maintaining the state's coral reefs and other tropical marine habitats. The current also provides energy to feed tropical storms and hurricanes and can transport pollutants, like spilled oil and marine debris, through the Gulf.

The sandy beaches of Florida's Gulf Coast and the Florida Keys provide critical nesting habitat for endangered shore birds, beach mice and sea turtle populations. Seagrasses and patch reefs off both Florida Bay and north Florida's Big Bend cover the sea floor for hundreds of square miles, providing food, important nurseries and shelter for incredible numbers and diversity of marine life and migratory waterfowl.

The state's ecosystems support not only thriving biological communities, but also a world-class tourism industry, providing many service-related and manufacturing jobs. The Everglades drains into the coastal estuaries of South Florida's Ten Thousand Islands and Florida Bay, supporting significant fisheries and wildlife that attract visitors from around the world and generate recreational and business opportunities for Floridians. Florida's Gulf Coast beaches are another example of ecosystem services of economic interest. They include the famed Emerald Coast along the Florida Panhandle and the beaches of Tampa Bay, Clearwater and St. Petersburg. Siesta Beach on Siesta Key, a barrier island southwest of Sarasota, was named the top beach in the country in 2011 by "Dr. Beach," producer of "The Best Beach in America." Each year Florida routinely has at least one beach on the top 10 list since its creation in 1991.[142]

Tourism was responsible in 2010 for welcoming more than 82.3 million visitors, who spent over $62.7 billion, generating 22 percent of the state's sales tax revenue and employing nearly a million Floridians.[143] Each year, almost $15 billion and 141,373 jobs result directly from fish and wildlife in Florida, and an additional nearly $17 billion and another 202,743 jobs are a result of boating activities in state waters.[144] Wildlife viewing is a significant pastime in Florida, accounting for $5.2 billion and 51,367 jobs in 2006. Also in 2006, almost 1.6 million people visiting the state participated in wildlife viewing in Florida, the majority of whom came to view coastal and marine wildlife.[145]

Military installations along Florida's Gulf Coast also provide significant economic activity for the state. Major military training and testing operations off northwest Florida include the Pensacola Naval Air Station and the Eglin, Hulbert Field and Tyndall Air Force Bases. These bases, along with the U.S. Central Command and Special Operations in Tampa, the Key West Naval Air Station, and the Homestead Air Force Base, produce significant economic and security benefits for the state and nation. Statewide, defense-related spending was $64.8 billion in 2010 and accounted for 686,181 direct and indirect jobs.[146] That spending is estimated to approach $67.7 billion in 2013.[147]

Gulf Coast ports at Pensacola, Panama City, Port St. Joe, St. Petersburg, Tampa, Port Manatee and Key West also add significant revenues and jobs to the state. These Gulf Coast ports account for about $10.6 billion per year in economic activity supporting 125,000 direct and indirect jobs.[148] The ports ship agricultural and industrial products and provide cruise ship access to customers and commerce throughout the wider Caribbean, in South America, and around the world. Florida is a top port of departure for cruises in the world, and more cruise ships are berthed here than anywhere else. Florida ports, including Tampa and Key West, provide popular day-long excursions too.

Maintaining healthy, sustainable fisheries is vitally important to the state, as recreational and commercial fishing contribute immensely to the economy, quality of life and character of Florida's coastal communities, particularly in the Gulf. Florida is one of the nation's premiere destinations for recreational fishing and is marketed as the "Fishing Capital of the World." Florida also has more world record fish catches than any other state or country.

Florida also leads all states in economic return for its marine recreational fisheries. Recreational saltwater fishing alone contributes over $5 billion and more than 50,000 jobs to the state's economy each year.[149] In 2008 to 2009, more than a million people—a third of whom were not Florida residents—bought marine recreational fishing licenses.[150] Also during this time, more than 3,400 for-hire fishing licenses were purchased.[151] In 2008, Florida Gulf Coast recreational anglers took 16.9 million trips: 9.6 million private/rental, 6.7 million by shore and 595,000 by party/charter boat.

The Department of Commerce has ranked Florida's commercial fishery as the second highest of all states for in-state sales at $13 billion annually, and the tenth highest in total landings revenue at $116 million annually. Florida is also the third highest state for jobs supported by commercial fishing, providing 64,744 jobs in 2009.[152]

Another important Florida fishery, particularly on the Gulf Coast, is oysters. Florida produces almost 13 percent of the Gulf's oyster catch (by weight), a $6.9 million annual dockside value to the state.[153] Apalachicola alone produces about 90 percent of Florida's oyster harvest.[154]

Florida's ecosystems also support new economic opportunities for fast evolving technological advancements in ocean energy, engineering, pharmaceutical, aquaculture and other marine related industries and jobs. Florida is home to at least 20 major public and private marine research and education facilities that help support these new economies for the future.[155]

State Priorities

The Task Force Strategy lays out four overarching goals. Florida's priority actions for each goal are described below.

Restore and Conserve Habitat

Florida has experienced major land use changes over the past 200 years, which have affected the state's diverse habitats. A great deal of development has occurred, particularly after World War II as people discovered Florida's sandy beaches, sunny climates and cheap land, which they modified to meet their particular needs. Agriculture, silviculture and community development required major changes in forest and wetland cover and drainage. Transportation corridors and fire cessation have resulted in habitat fragmentation and alterations. Ditching and draining, shoreline hardening, runoff and excess sediments all contributed to major losses of productive coastal and some marine habitat.

Florida also has a long history of land acquisition for preservation, protection and restoration of natural resources, which has resulted in an extensive network of protected critical natural habitat throughout the state. Of specific note is the extensive level of federal and state coordination and resource commitment to South Florida ecosystem restoration through the CERP.

Florida's priority actions for habitat conservation and restoration include:

- Protect, stabilize and restore salt marsh, seagrass, oyster, coral reef, beach, dune mangrove and other important marine bottom habitat in strategic locations where human-made and storm impacts have occurred or are likely to occur in the future.
- Partner with Florida Gulf Coast military installations to integrate their ongoing conservation strategies and land acquisition efforts for military readiness with Florida's ecosystem restoration and protection goals.
- Implement policy and program changes necessary to reestablish natural, historical water flow (quality, quantity, timing and distribution) on the Apalachicola, Suwannee, Peace and Caloosahatchee Rivers; Florida Bay; and other riverine and estuarine systems connecting to the Gulf.
- Strategically acquire, buffer and protect identified properties in state and federal acquisition programs to provide watershed improvements and wildlife corridors to downstream estuarine and marine habitats of ecological and economic importance to the Gulf.
- Work with federal and state partners to establish a financial and public policy process for the beneficial use of sediments and rock from major dredging projects in restoring barrier island, beach, hard bottom and other coastal habitats.
- Continue to work with federal partners, private landowners and stakeholders to implement the CERP where it will improve and protect Gulf ecosystems in Southwest Florida and the Keys.
- Refine and implement inlet management plans to restore the natural flow of sediments around inlets to down drift beaches.
- Develop and implement invasive species' eradication and management plans to address impacts to natural ecosystems within the Gulf region.
- Restore and manage critically eroded sandy beaches and dunes for upland protection, recreation, tourism and wildlife.
- Work with federal and local agencies to develop and implement a hard bottom mitigation strategy to restore habitats where damage from storms, transportation and navigation projects, beach restoration, and other impacts from public projects have occurred or will occur.

FLORIDA FOREVER

Florida has the largest public land acquisition program of its kind in the United States —started as the Preservation 2000 program in the late 1980s, followed by the Florida Forever program—which could be used as a model for coastal habitat protection in other Gulf states and the nation. Florida has approximately 9.9 million acres of federal, state and local conservation land (over a quarter of the state); more than 2.5 million acres were purchased under the Florida Forever and Preservation 2000 programs. Since July 2001, Florida Forever has acquired more than 673,753 acres of land, worth $2.83 billion.

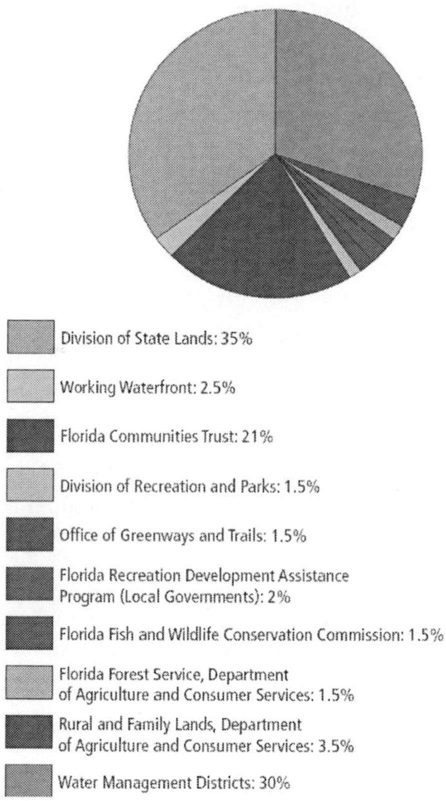

Florida Forever Funding Distribution

The Florida Department of Environmental Protection distributes Florida Forever funding to a number of state agencies and programs to purchase public lands in the form of parks, trails, forests, wildlife management areas and more. All of these lands are held in trust for the citizens of Florida.

Restore Water Quality

While Florida has some of the clearest and cleanest waters in the Gulf of Mexico, as of 2010 more than 1,300 square miles of the state's estuaries are deemed "impaired" for pollutants other than mercury and therefore not fully meeting their designates uses.[156] Mercury impairments predominantly are caused by atmospheric deposition. In addition to developing a comprehensive statewide Total Maximum Daily Load (TMDL) for mercury, Florida is especially focused on the issue of excess nutrient pollution, and has one of the most robust water quality protection programs in the country.

In addition, surface water and groundwater systems are closely interconnected throughout the state. Therefore, in addition to maintaining and restoring the water quality of estuaries and coastal waters, and the freshwater rivers and streams that flow into them, it is vital to protect the groundwater sources within aquifers.

A growing Florida presents constant challenges in protection of water resources. The need and desire to increase the use of the Gulf Coast ports and other coastal and inland water resources, along with a growing population, increases competition for Florida's water resources. Challenges include cleaning up historical contamination associated with coastal industrial areas, managing the influence on water quality of active ports, and addressing urbanized estuaries near Pensacola, Tampa, Naples and Fort Myers.

Setting Nutrient Limits for Florida Estuarine and Coastal Waters

The State of Florida began a nationally recognized water quality improvement program with the passage of the Watershed Restoration Act of 1999 (s. 403.067, F. S.), which directs the Department of Environmental Protection to scientifically evaluate the quality of Florida's surface waters and promote the mechanisms necessary to clean up pollution. The Act was created specifically to implement the federal TMDL program, which is a systematic approach to establishing how much pollution water bodies can assimilate while remaining suitable for public uses and meeting quality standards. The Department is working closely with Florida stakeholders to set numeric limits on the amount of nutrients (nitrogen and/or phosphorus) that can be discharged into the state's estuarine and coastal waterways to better protect these waters from the adverse effects of excess nutrient enrichment. The established limits, if attained, ensure that the designated uses of Florida's marine waters are maintained.

The primary purpose of numeric nutrient criteria is to protect healthy, well-balanced natural populations of flora and fauna from the effects of excess nutrient enrichment. Implementation of these criteria can prevent over-enrichment, and can be used to identify waters impaired by nutrients in need of restoration. The criteria would also support full recreational use of the state's marine waters.

One of the most visible impacts of nutrient (nitrogen and phosphorus) enrichment in Florida is freshwater HABs. Understanding the causes of HABs and ways to mitigate them is vital to protecting the downstream estuaries. One of the solutions is beneficial reuse of wastewater in Florida. Although a large percentage of Florida's population uses advanced community collection and wastewater treatment systems, leading to a significant amount of beneficial reuse, a number of coastal communities still use septic tanks or older and inadequate collection and treatment systems, which do not allow beneficial reuse.

In Florida, Gulf HABs are commonly known as red tides, which are a high concentration of naturally occurring, toxic, microscopic marine algae, originating in offshore waters and concentrating in coastal waters. Red tides can kill fish and other marine life, contaminate shellfish, and be dangerous to people as well. Massive blooms can also have a devastating impact to the tourism-based economy when fishing areas are closed and breathing becomes difficult for beach goers due to toxic sea spray along the coast. The exact cause and mechanism for this phenomenon are unknown.

Florida's priority actions for restoring and protecting water quality include:

- Reduce excessive sediment, pollutant and nutrient loads in waterways by improving the management of urban and agricultural stormwater runoff, increasing the treatment levels of wastewater and encouraging their reuse, restoring inland waterways that flow to the Gulf, restoring canals and altered floodplains to represent more natural systems, and promoting more sustainable adjacent land use policies.
- Improve education and incentives related to nonpoint source pollution (agricultural, residential and urban) and promote local BMPs and alternatives.
- Focus water quality and clarity improvements to best promote seagrass, oyster and coral restoration.
- Improve understanding of the sources, bioaccumulation and effects of toxic chemicals (such as pesticides, dioxins and PCBs), metals (such as mercury) and other environmental contaminants, as well as how to cycle these pollutants out of sediments and nearshore waters. Improve understanding of the ecological harm of increasing acidification in Gulf waters and identify ways to reverse this trend.
- Continue to monitor, investigate, and possibly mitigate and ameliorate HABs off Florida's Gulf Coast.
- Evaluate the feasibility of removing or improving control structures that are impeding tidal exchange to estuaries to restore more natural salinity conditions that serve as primary nursery areas.

Replenish and Protect Living Coastal and Marine Resources

Habitat loss and degradation have led to declining wildlife populations in coastal areas. These natural areas are critical to many species, including shorebirds, sea turtles, marine fish, beach mice and marine mammals. In addition, they are vital to tourism, serving as sites that many coastal visitors enjoy for recreation.

Increasing levels of disturbance in coastal areas from beach nourishment, coastal development and recreational activities, as well as the implications of sea-level rise, are all challenges facing coastal and marine fish and wildlife and other resources. An integrated approach among all stakeholders that focuses on fish, wildlife and habitat needs, as well as socioeconomic issues, is greatly needed to address activities that affect coastal and marine resources.

Florida's priority actions for replenishing and protecting living coastal and marine resources include:

- Develop and implement management plans for conserving threatened, endangered, and other protected species and maintaining commercial and recreationally important species. Management plans should include ways to improve monitoring of coastal and marine resources; improve research into those resources; minimize adverse impacts from human activities; maintain sustainable native populations; and protect, restore and maintain critical habitat for listed fish, wildlife and plants in coastal areas.

- Involve the public in developing initiatives to help educate citizens and communities on the importance of coastal wildlife conservation, shorebird protection and sea turtle monitoring.
- Develop and implement programs to balance and integrate the interests and needs of people living and recreating in coastal areas with the needs of fish and wildlife species dependent on marine and coastal habitats.
- Develop large-scale strategies for seagrass, oyster and coral reef restoration and fish hatcheries and aquaculture programs for marine species propagation and enhancement.
- Develop and implement invasive species eradication and management plans to address impacts to natural ecosystems within the Gulf region.
- Develop coral propagation and post-storm and vessel grounding coral restoration plans.

Enhance Community Resilience

Florida's coastal resources face substantial threat from red tides, coastal wildfires, population growth, development, and other natural and human-made hazards. From 1960 to 2009, 63 major disasters were declared in the state, mostly due to hurricanes, tropical storms and coastal flooding.[157] In addition, projected sea-level rise threatens to exacerbate the vulnerability of Florida's at-risk coastal resources. These significant coastal resources often provide the first line of defense from a natural disaster and are, therefore, important to the overall resiliency of the state.

Florida has already taken steps to address community resiliency through projects aimed at post-disaster redevelopment and sea-level rise adaptation. The state has also begun integrating hazard mitigation and community resiliency into the local comprehensive planning process. However, with 80 percent of the state's population living in coastal areas, additional efforts will likely be needed to protect these communities.

The state's priorities for community resiliency include:

- Compile, review and summarize "coastal elements" of Florida's 23 Gulf Coast counties' Comprehensive Growth Management Plans for continuity and consistency in natural resource and community infrastructure protection to aid in Gulf restoration and community resiliency.
- Incorporate projects to address natural resource protection into local coastal hazard mitigation plans.
- Promote the development of Community Conservation Plans that include proven elements of coastal construction and conservation, emergency response and risk reduction, and economic development.
- Improve knowledge of the economic value of environmental services provided by Gulf resources in terms of long-term community sustainability, growth and resilience.
- Update the Spill Response Contingency Plans through a multi-disciplinary process to incorporate better preparation, local involvement, and updated ecological data.

Promote wider understanding of the plans and involvement among potentially affected parties.

- Support local efforts to develop redevelopment plans that address natural resource and economic protection and recovery following a disaster and that support statewide, long-term recovery.
- Coordinate statewide efforts to address sea-level rise, saltwater intrusion and other impacts from climate change for both the natural and built environment, especially potable water sources.

Louisiana

Background

Coastal Louisiana is one of the most productive estuaries in the world.[158] The state's coastal area is composed of the largest deltaic plain in North America and the world's seventh largest delta, tens of thousands of square miles,[159] and more than 7,700 miles of tidal shorelines.[160] This area also plays a critical role in the nation's energy and economic security. South Louisiana is the top source of offshore energy in the country,[161] the largest source of wild seafood in the lower 48 states, and home to five of the top 15 ports in the nation.[162]

Southern Louisiana is America's Wetland, a landscape created by the Mississippi River over the last several thousand years. This land-building occurred as the river periodically switched course to find a more efficient path to the Gulf and spread its sediment, nutrients and freshwater across southeast Louisiana. This process also created the Chenier Plain of southwest Louisiana, a unique area of alternating wetlands and oak ridges from which the area gets its name. Louisiana and much of its adjacent Gulf Coast was literally built by the Mississippi River system.

Since the arrival of Europeans, people have tried to manage the river for their own safety and prosperity. An unintended result of these repeated attempts to manage the river, combined with other factors over the past three centuries, has been significant losses of wetlands and coastal habitat with nearly 2,000 square miles of wetlands lost since the 1930s.[163] The practices of constructing levees, straightening, channelizing, dredging and otherwise managing the river have succeeded in preventing riverine floods, facilitating economic development, and maintaining a deep-draft navigation channel, but have also separated the river from the adjacent lands and wetlands—its deltaic plain. The valuable sediment, nutrients and freshwater resources that once created and sustained Louisiana's and much of the northern Gulf's coastal ecosystem are lost to the deep waters of the Gulf of Mexico. Without reconnecting the river system to its deltaic plain, Louisiana's coast will remain unsustainable and the Gulf ecosystem will degrade further in turn.

Still, Louisiana has the largest expanse of coastal wetlands in the lower 48 states.[164] More than 5,600 square miles of coastal swamp and marsh are present in coastal Louisiana,[165] an area larger than the state of Connecticut. This coastal region is known as a working coast. In addition to having extraordinary ecological diversity and productivity, it encompasses energy resources and infrastructure, national and international commerce, fisheries and wildlife habitat, natural storm protection and water quality enhancements, and a rich cultural heritage.

As the land loss crisis continues in Louisiana, and the Gulf of Mexico encroaches on Louisiana's coastal communities, the vulnerability of this region to damage from tropical storms increases every day. While coastal communities are at risk, so are billions of dollars of infrastructure funded by industry, municipalities, the state, and the United States. Roads and highways, utilities, railways, ports, and navigation systems will be lost and more than a million citizens will be forced to move. In 2005, Congress made a supplemental appropriation of $94.8 billion to respond to the impacts of Hurricanes Katrina, Rita and Wilma.[166] In order to protect this investment and prevent the need for additional large-scale investments in responses to catastrophic events, immediate and meaningful action must be taken to achieve a sustainable coastal Louisiana and Gulf Coast.

Energy infrastructure. The oil and gas industry has established and concentrated coastal and offshore exploration, production, transportation infrastructure, and refining capacity along Louisiana's Gulf Coast. Nearly 125,000 miles of oil and natural gas pipelines pass through the marshes of coastal Louisiana, and throughout the state and its offshore area in the Gulf of Mexico.[167] The Henry Hub in Erath, Louisiana, is the pricing point for natural gas in North America[168] and the nexus of nine interstate and four intrastate pipelines,[169] and Port Fourchon provides a port and supply point for 90 percent of the offshore drilling operations in the Gulf of Mexico.[170] The network of energy facilities in and around Louisiana's Gulf Coast produces or transports nearly one-third of the nation's oil and gas supply and is tied to 50 percent of the nation's refining capacity.[171] Louisiana's coast also contains the Louisiana Offshore Oil Port, the only port in the nation capable of offloading supertankers;[172] about 38,000 oil and gas wells;[173] and two of four storage sites for the Strategic Petroleum Reserve nationwide.[174] Barrier islands and wetlands help buffer this infrastructure from storm damages.

National and international commerce. Ten major navigation routes, including the Mississippi River, are located in coastal Louisiana. Five of the 15 busiest ports in the United States, ranked by total tons,[175] are also located in this region, handling nearly 450 million tons of waterborne cargo each year.[176] This represents 20 percent of annual U.S. waterborne commerce.[177] Each year, the Port of South Louisiana and the Port of New Orleans together account for $150 billion and nearly 20 percent of the U.S. import/export cargo traffic.[178] The state's top three ports handle an estimated 55 to 70 percent of all U.S. exported corn, soy and wheat. Louisiana's ports also handle other commodities such as steel, coffee, rubber and chemicals.[179] Without barrier islands and wetlands, all of this infrastructure would be at greater risk when storms come ashore.

Fisheries and wildlife habitat. This ecosystem is the nation's largest shrimp, oyster, crawfish and blue crab producer and provides 26 percent (by weight) of the commercial fish landings in the continental United States.[180] In fact, Louisiana is second only to Alaska in annual volume of seafood landings,[181] and three of the nation's top seafood ports by volume are in Louisiana.[182] These resources are processed and shipped throughout the world, providing jobs for almost 30,000 Louisiana citizens, as well as jobs in other states.[183]

The Mississippi Flyway passes directly over south Louisiana, and more than five million migratory waterfowl pass through or spend the winter in Louisiana's marshes.[184] In addition, the coastal landscape provides stopover habitat for millions of neotropical migratory birds on their journeys across the Gulf of Mexico. Hundreds of fish and wildlife species, as well as the jobs and recreational opportunities associated with birding, hunting, fishing and eco-tourism, all depend on the barrier islands and wetlands along Louisiana's Gulf Coast. Sixteen endangered, threatened or otherwise protected species are found in south Louisiana—including the Gulf sturgeon, Louisiana black bear, and several species of sea turtles—as well as the recently recovered bald eagle.[185]

Water quality. Nutrients from farms and other sources in the Mississippi River watershed cause the largest low-oxygen (or hypoxic) conditions off the coast of the United States. The "dead zone" is virtually void of marine life and has averaged over 6,000 square miles since 1990.[186] High concentrations of nutrients in the northern Gulf of Mexico contribute to the growing problem of hypoxia, or low-oxygen conditions, in offshore coastal waters. As increasing amounts of river water are diverted into marshes as part of restoration projects, these nutrients will help sustain wetland plants or be processed in the soil, rather than contributing to a nationally significant water quality problem.

Culture. People have lived in south Louisiana for over 12,000 years, using the abundance of the rivers and Gulf Coast to extract resources and facilitate trade. When New Orleans was founded nearly 300 years ago, it quickly became a center of international commerce, attracting people from around the world. These diverse peoples lived in proximity while retaining their own identities, a trend that defied typical melting pot dynamics and created a multi-faceted culture that endures today. The Native American Chitimacha people have lived on Louisiana's coast for at least 2,500 years and state as part of their beliefs, "We have always been here." The regard for land and family expressed by this sentiment is one that many residents of south Louisiana share.[187] In fact, according to the 2010 Census, Louisiana has the highest percentage of native-born residents (81.9 percent) of any state in the nation.[188]

Coastal Crisis

In spite of these attributes, Louisiana's Gulf Coast and its vital natural resources are greatly imperiled. The effects of four major hurricanes (Katrina, Rita, Gustav and Ike) within a four-year period (2004 to 2008) resulted in a cumulative net loss of 849.5 square kilometers (328 square miles), an amount equivalent to decades of coastal land loss during periods of low storm frequency.[189] Louisiana's coastal crisis was further compounded in 2010 by the Deepwater Horizon oil spill, which released millions of barrels of oil into the Gulf, resulting in the closure of state and federal waters to fishing and affecting thousands of miles of shoreline, bayous and bays in Louisiana and other Gulf Coast states. With more of its precious coastal assets lost each year, the state and nation are under increased pressure to identify new tools and resources to protect and restore coastal environments, assets and communities in a resilient manner to ensure that Louisiana's vital coastal resources are better equipped to withstand future threats.

LOUISIANA COASTAL PROTECTION AND RESTORATION PLAN

Following Hurricanes Katrina and Rita, Congress directed the U.S. Army Corps of Engineers to develop a comprehensive ecosystem restoration and hurricane protection plan for coastal Louisiana (the Louisiana Coastal Protection and Restoration plan, or LACPR) and Mississippi MsCIP. The State of Louisiana is now working with the Corps to further inform project prioritization through the development of the State Master Plan.

The Coastal Protection and Restoration Authority of Louisiana has directed the investment of over $1.7 billion in state and federal funds over the last three years in ecosystem restoration and community sustainability measures and improvements to coastal resource management.[190] Recent reports have indicated that these investments, together with system resilience, have resulted in coastal land gains in some areas; however, an appropriately balanced approach among investment priorities (regardless of the type of action or party involved) is needed to achieve sustainable management of the coast.

Master Plan

In 2007, the state developed the Master Plan for a Sustainable Coastal Louisiana in response to the Louisiana legislature's directive to integrate coastal protection and restoration activities, establishing four broad planning objectives as benchmarks for implementing coastal protection and restoration projects. The 2007 Master Plan also identified large-scale measures needed to achieve a sustainable coast. This initial Master Plan, though groundbreaking, was a conceptual document (which helped to inform LACPR) and was not intended to address all of the numerous complex issues that coastal Louisiana faces. To accommodate the dynamic nature of coastal processes, the Coastal Protection and Restoration Authority of Louisiana is working to update of the Master Plan every five years to incorporate new data and planning tools as they become available. The first update of the Master Plan must be submitted to the Louisiana legislature in March 2012. The State of Louisiana's 2012 Master Plan, which is supported by multiple federal agencies, will present a new approach for considering the future of the coast. This approach will identify protection and restoration projects that achieve multiple objectives (integrated planning). The plan will also propose a specific order for building projects to ensure that the most important projects are constructed first (prioritization). The state began laying the foundation for this new approach in 2010 with the development of a Prioritization Tool designed to help identify portfolios of "high-value" projects. Progress on the Master Plan Update has continued in 2011 with the establishment of a Master Plan Delivery Team that is committed full-time to the development of the 2012 update.

Louisiana's 2012 Master Plan will provide a pathway to sustainability. The plan will answer two questions that coastal residents have long been asking: just how bad is the coastal crisis going to get, and what will be done about it? People are worried about flooding, and they want to know what to expect. The plan will describe solutions to Louisiana's land loss and flooding problems that affect people, businesses and natural areas. This will help communities and their leaders know what to expect as they plan for the future.

LOUISIANA'S COASTAL RESOURCES: WHAT'S AT STAKE

Coastal Louisiana currently contains about 40 percent of the coastal marsh in the continental United States, but accounts for about 90 percent of coastal wetland loss in that region.[191] Louisiana's coastal crisis is not merely a state concern, but a national concern. At stake are resources and industries that support millions of jobs nationwide, including:

- The highest rate of crude oil production and second highest rate of natural gas production in the nation (includes Outer Continental Shelf production).[192]
- Eighteen active oil refineries. The operating refineries in Louisiana account for approximately 19 percent of the nation's refining capacity, ranked second in the nation,[193] supported $70.2 billion in sales and $12.7 billion in household earnings, and supported 320,280 jobs in 2005.[194]
- A chemical industry that generates nearly $1 billion each year in tax revenues for Louisiana and billions of dollars to the nation's economy. The Louisiana chemical industry directly employs over 30,000 citizens, paying salaries that total over $1.7 billion per year.[195]
- Nearly 450 million tons of waterborne commerce (20 percent of all waterborne commerce in the nation).[196]
- The largest port in the nation in terms of total tonnage shipped (the Port of South Louisiana).[197]
- Five of the 15 largest ports in the nation.[198]
- Nearly 3,000 miles of deep and shallow draft navigation channels to support waterborne commerce.[199]
- Approximately 26 percent of the total commercial fishing catch in the lower 48 states ($1 billion annually).[200]
- A robust recreational fishing industry with nearly $2 billion in annual revenues.[201]
- The largest fur harvest in the nation.[202]
- More than five million migratory waterfowl that pass through or spend the winter in Louisiana's marshes.[203]
- Eleven National Wildlife Refuges covering an area greater than 524 square miles.[204,205]
- Nearly two million residents living in parishes within the state's coastal zone.[206]
- An estimated 25 million domestic and international visitors in 2010.[207]
- $9.3 billion added to the economy through tourist spending in 2010.[208]

People also want action. They want to be protected from flooding and they want wetland loss addressed. The 2012 Master Plan is focused on taking the action needed to address these concerns. To deliver an effective but feasible action plan, the 2012 Master Plan will propose projects that will reduce flooding risks using a range of methods, from levees to building up landscape features to helping communities flood-proof their homes. The plan will also propose projects that build wetlands, restore coastal landscape features, and help sustain the multitude of ecosystem services that support communities, nationally significant industries, and commercial operations. In order to bring flood risk reduction and wetland restoration together into one integrated plan, the 2012 Master Plan will present portfolios of projects that combine both objectives. The 2012 Master Plan will present the best possible use of dollars based on the current state of knowledge, while laying the groundwork for improvement in coming years.

State Priorities

The Task Force Strategy lays out four overarching goals. The State of Louisiana's priority actions for each goal are described below.

Restore and Conserve Habitat

- *Improve river and federal navigation channel management.* Current management practices have effectively separated the Mississippi River from its floodplain, causing or increasing land loss, nutrient loading to the Gulf and salt water intrusion. Current management practices have resulted in an unsustainable coastal and Gulf of Mexico ecosystem as well as navigation and flood damage risk reduction systems.
  Needed actions:
  - Establish sediment delivery targets to restore coastal wetlands.
  - Mandatory use of dredged sediment for habitat restoration to mitigate negative effects of navigation channel management.
  - Evaluate lower river management holistically.
    - Construct demonstration projects that evaluate alternative management techniques:
      - Sediment traps.
      - Permanent sediment transport infrastructure.
      - Siphons and wetlands for nutrient filtering.
  - Implement river reintroduction projects that mimic or enhance natural hydrologic processes and provide freshwater, sediments and nutrients to help sustain or restore wetlands.
  - Launch a pilot water quality credit program that evaluates wetlands capacity to filter nutrients.
  - Pursue embankment stabilization and shoreline protection measures to maintain federal navigation channels within authorized dimensions.
- *Maximize and stabilize revenue streams.* Current restoration programs (e.g., the Coastal Wetlands Planning, Protection and Restoration Act [CWPPRA], the Coastal Impact Assistance Program [CIAP], State Surplus, the Water Resources Development Act) use a project-by-project planning and implementation approach, limiting capabilities of the overall restoration program.
  Needed actions:
  - Establish a multi-year budget, coordinated among federal agencies, that would allow for a programmatic approach to ecosystem restoration.
  - Issue regulations and projected revenue streams associated with the Gulf of Mexico Energy Security Act.

    - Facilitate and support the development of "ecosystem services credits" and other innovative funding sources for coastal restoration (carbon credit, water quality credits, natural resources damage credits, etc.).
    - Develop options to use mitigation funds and the "pooling" of resources from separate projects which may require changes to existing regulations on mitigation.
- *Expedite project implementation and remove barriers to large-scale restoration.* It has often taken many years to implement projects through the existing water resources planning and implementation process. It is necessary to accelerate the process, especially when dealing with a crisis like the one coastal Louisiana is facing.
  Needed actions:
    - Reevaluate the definition and application of the federal standard for the beneficial use of dredged materials and how it relates to the requirements of the Coastal Zone Management Act.
    - Improve/expedite the process for acquiring sand leases in federal waters.
    - Remove barriers to effectively crediting local cost share of projects.
    - Focus on mitigation for the negative environmental impacts caused by the ongoing operation and maintenance of flood control, navigation and other projects.
    - Use CWPPRA or a similar multi-agency approach for the development and implementation of large-scale projects.
    - Explore modifications to the National Environmental Policy Act process for ecological restoration projects within existing legal authorities.

Restore Water Quality

- *Reduce hypoxia.* Improve water quality by reducing excess nutrients flowing into, and hypoxic conditions in, the Gulf of Mexico.
  Needed actions:
    - Reconstruct distributaries to mimic historic deltaic processes, sustain or restore wetlands, and filter excess nutrients from water entering the Gulf.
    - Launch a pilot water quality credit program that evaluates the effectiveness of the distributaries and associated wetlands to filter nutrients.
    - Execute cooperative endeavors with upper-basin states to construct wetlands buffers and other measures to reduce nutrient inputs in the Mississippi River system.

Protect and Replenish Living Coastal Marine Resources

- *Restore historical oyster reefs.* The Louisiana coast was once the location of massive offshore oyster reefs in Bay Gardene and Barataria, Terrebonne, Atchafalaya and Vermilion Bays. The state has a goal of establishing more acres of harvestable oyster reefs and living shorelines than ever before.
  Needed actions:

    - Manage reefs to support a suite of ecosystem services, including sustainable harvest, fish production, water filtration, nitrogen removal and protection of shorelines and marsh land.
    - Assess and implement Gulf-wide restoration of near-shore habitats that promote shoreline stability and habitat diversity, such as oyster reefs, mangroves, marshes and beach/dune systems.
    - Establish buffers landward of marsh shorelines that are under pressure from shoreline development and sea-level rise to allow for marshes to migrate.
    - Integrate living shoreline concepts into comprehensive restoration and protection efforts.
- *Eliminate invasive species.* Louisiana is home to dozens of invasive species that destroy the precious wetlands we are trying to preserve and restore, displace naturally occurring species and reduce habitat value to our native wildlife and fish.
  Needed action:
    - Develop and implement invasive species management and eradication plans. Use existing state and regional partnership models such as the CWPPRA Coast-Wide Nutria Control Program to guide implementation.

Enhance Community Resilience

- *Enhance coastal community resilience.* Use comprehensive, scientifically based, and stakeholder-engaged coastal improvement plans at the state and local levels such as the 2012 Master Plan for a Sustainable Coastal Louisiana.
  Needed actions:
    - Support and expedite implementation of community resilience features of the 2012 Master Plan for a Sustainable Coastal Louisiana.
    - Expedite decision making for federally authorized hurricane protection, flood control and ecosystem restoration efforts.
- *Establish alternative project delivery processes.* A new project development and implementation process must be established that can respond to the urgency in the Gulf. The recovery of 2005 hurricanes alone cost taxpayers an estimated $150 billion. Some of these vulnerabilities continue to exist today. The state-federal CWPPRA, the Gulf of Mexico Energy Security Act, the revised CIAP and some post-hurricane recovery programs—including alternative environmental compliance processes—are models that can be evaluated to develop this new process. Local, state and federal governments cannot afford to allow this level of risk to continue.

Mississippi

Background

If the best things come in small packages, Mississippi's coast may be the diamond of the Gulf. The state packs a lot of natural beauty, ecological and social diversity, culture, cuisine and entertainment into a small area, which helps drive one of the most important economic engines in the state.

Along the coast, Mississippi spans roughly 69 miles between the Alabama and Louisiana state lines.[209] The state's three counties that border the Gulf Coast, Hancock, Harrison and Jackson, have a population of 370,702, compared to a statewide population of 2,967,297.[210]

The tidal shoreline in this area is 359 miles and encompasses barrier islands, mainland coast, bays, lagoons and river shorelines.[211] Mississippi's coastal waters also include 758 square miles of large estuaries, smaller bays and tidal rivers, creeks, and bayous.[212]

The Mississippi Coastal Plain is laced with scenic streams, both wild and lazy, from the almost unpronounceable Tchoutacabouffa River to the longest undammed river in the lower 48 states, the Pascagoula. These streams are typically shallow and clear blackwater streams, with moderate flow and wide sandbars, gradually becoming wider, deeper and more sluggish as they flow toward the Gulf.

The streams and estuaries of this area are home to a remarkable variety of plants and animals. A large number of threatened and endangered species are found here in what is recognized as one of the most biologically diverse regions in North America. The area is also known for the large number of amphibian, reptile and bird species, and ranks in the top 10 for endemic species of reptiles, amphibians, butterflies and mammals.[213] The beauty of the landscape and the habitats that make this area so popular for outdoor recreation also provide safe haven for the plants and animals that live there. Good water quality and protection of habitat are essential to support these species.

The Grand Bay National Estuarine Research Reserve

The Grand Bay National Estuarine Research Reserve straddles the Mississippi/Alabama state line. This 18,000-acre reserve represents one of the largest, relatively undisturbed estuarine marsh/pine savanna habitats remaining along the northern Gulf of Mexico. Diverse habitats such as salt pannes, saltwater and freshwater marshes, bayous, oyster reefs, and seagrass beds provide critical habitats for many of the region's important commercial and recreational species of fish and migratory birds. These habitats serve as nursery areas as well as breeding and feeding grounds for shrimp, red drum, speckled trout, oysters, Wilson's plovers, peregrine falcons, Alabama red-bellied turtles and other species of concern.

Multiple events have shaped the coastal environment, changing its topography, altering the bottom substrate of the estuaries and disrupting the lives of its residents, including humans and aquatic and terrestrial organisms. In August 1969, Hurricane Camille made landfall in Harrison County, devastating coastal Mississippi with tremendous economic damage and loss of life. This powerful storm eroded the offshore barrier islands and cut Ship Island into two separate islands. In addition, it increased the coast's susceptibility to future storm events. This increased susceptibility manifested itself in August 2005, when Hurricane Katrina made landfall along the Louisiana-Mississippi state line. The accompanying storm surge wreaked havoc on the area, destroying homes, businesses and coastal habitat.

Most recently, in April 2010, the *Deepwater Horizon* platform exploded, leading to the worst oil spill in U.S. history. The spill provided an intense focus on the Gulf, as well as renewed momentum to continue the development of far-sighted strategies to improve the Gulf

Coast's ecosystems and economy. Mississippi Governor Haley Barbour appointed the Mississippi Gulf Coast Commission in August 2010, to draft a framework for a coastal restoration program. The Commission, composed of leaders from government, private business, academia and nonprofit organizations, collaboratively produced *Vision for Gulf Coast Recovery, Restoration and Protection*, which was submitted to Secretary of the Navy Ray Mabus.

THE MISSISSIPPI BARRIER ISLANDS

The Mississippi Barrier Islands dangle off the mainland like a string of pearls in the blue-green waters of the Gulf of Mexico. Petit Bois, Horn, East Ship, West Ship and Cat Islands are mostly owned by the National Park Service and are part of the Gulf Islands National Seashore. Accessible only by boat and virtually untouched by human development, these islands are a great place to view nesting ospreys, and they provide habitat for many shore birds and wildlife species, including the American alligator and piping plover. The islands also serve as a natural barrier, buffering the mainland from storms and from the *Deepwater Horizon* oil spill. These islands have had their lifeline of renourishing sand and sediment severed by navigation channels that have interrupted the natural flow of sand to them, making them prime candidates for restoration efforts.

Mississippi coastal shorelines have been eroding steadily over the past several thousand years, driven primarily by relative sea-level rise, wave action, tropical storms/hurricanes and disruptions in the sediment transport system. From the 1950s to the 1990s, coastal marsh in Mississippi declined from an estimated 67,000 acres to 58,000 acres, a loss of 9,000 acres.[214] The loss of coastal wetlands continues today at rates similar to the historical trends; however, increased development pressure and expected increases in sea-level rise will likely accelerate the rate of loss.

The health and resiliency of the state's ecosystems and the economy are tightly woven together. Commercial and recreational fishing, gaming, tourism, energy production, manufacturing and shipping provide the base of a diverse and vibrant coastal economy. Commercial fishing and seafood processing is a natural extension of life in this coastal area. Hundreds of fishing boats make their home in Mississippi ports, and the seafood they catch generates thousands of business opportunities.

Tourism is driven in part by the seafood industry, as recreational fishing opportunities bring many visitors to Mississippi. The Mississippi Department of Marine Resources has built 200 acres of inshore reefs and approximately 16,000 acres of offshore reefs through its artificial reef program.[215] These reefs provide improved habitat for many bottom-dwelling species, like red snapper and grouper, and increase fishing opportunities for the public.

The Mississippi Gulf Coast has become one of the top gaming destinations in the country, with world-class entertainment, hotels and cuisine. The success of the Mississippi Gulf Coast tourism industry has a direct impact on the revenues generated for the state. Tourism on the Gulf Coast accounts for about $1.7 billion in visitor expenditures, 32 percent of state travel and tourism tax revenues and 23,000 direct jobs.[216] The Mississippi coast also offers attractive beaches, championship golf courses, museums and art galleries to complement its gaming and fishing industries.

Offshore oil and gas exploration and development further boost the area's economy. Providing experienced offshore workers, necessary logistical support and industrial capacity to generate that support are all-important sources of fuel for the coast's economic engine.

The largest cities on the Mississippi coast are Gulfport, Biloxi and Pascagoula. The Mississippi State Port at Gulfport is a hub of national and international commerce. It is an economic force for jobs and business activity alike. The port generates more than 2,000 jobs for Mississippi residents, with that number projected to increase to almost 5,500 with planned improvements and expansion as the city undergoes a comprehensive recovery from Hurricane Katrina.[217] Gulfport also lays claim to hosting "The World's Largest Fishing Rodeo" each Fourth of July, where anglers and sightseers flock each year to show off their catch and to marvel at the huge blue marlin, eerie king snake eel and other creatures from the deep that are brought to the scales. Biloxi is the home of Keesler Air Force Base, which houses the 81st Training Wing, the Air Force's largest Technical Training Group. The 81st Training Wing is the Air Force's computer and electronics training "Center of Excellence," which trained 22,925 airmen, soldiers, sailors and Marines in 2008.[218] Pascagoula is proud to be the home port of the largest military shipbuilder in the United States, having built over 70 percent of the existing U.S. Navy fleet. This shipyard is also the largest private employer in the state, providing approximately 11,000 jobs for residents of the northern Gulf region.[219]

Quaint small towns, like Bay St. Louis (on the western shore of the Bay of St. Louis) and Ocean Springs (at the mouth of Biloxi Bay), dot the Mississippi coast. These small communities offer interesting shops, art galleries and great local restaurants featuring fresh Gulf seafood. They epitomize Mississippi hospitality and attract visitors from all over the world.

State Priorities

The Task Force Strategy lays out four overarching goals. Mississippi's priority actions for each goal are described below.

Restore and Conserve Habitat

The Gulf will continue to have destructive storms and other disasters. The State of Mississippi must work to reduce the damage from these events, thereby protecting lives and homes, lowering repair and rebuilding costs, and improving the region's ability to recover quickly following a disaster. In partnership with the environment, the state can build capacity to protect and restore the coast's ecosystems from both natural and human-made disasters. Coastal communities can be better prepared for changes of all kinds by assessing risks in advance of those changes and making appropriate choices to mitigate harm.

Mississippi's key priorities for conserving and restoring habitat include:

- Conserve and protect the state's existing natural treasures.
- Restore the Gulf barrier islands and wetlands, including the development of a Sediment Management Master Plan to effectively use dredged materials and other sediment for restoration projects.
- Establish restoration and enhancement projects based on continuing investigation and assessment to ensure the renewal and long-term sustainability of fish and shellfish populations and their habitats, including marshes, seagrass beds and oyster reefs.

- Initiate community-based efforts to increase awareness of the importance of coastal resources and the BMPs to support conservation and renewal of these invaluable assets.

Restore Water Quality

It is imperative that the Gulf Coast have clean surface waters, safe drinking water and safe seafood. Good water quality is a critical component of coastal living. Not only is clean water essential for the health of the plants and animals that inhabit coastal waters, but it also supports jobs for those involved in fishing, shipbuilding and tourism; provides recreational opportunities for residents and visitors alike; and supports the local cuisine and lifestyle.

Nutrients, especially nitrogen and phosphorus, are an important water quality concern in the Gulf. The right balance of the kinds and amounts of nutrients carried into the state's estuaries and coastal waters is an important factor affecting the health of Gulf estuaries and open waters. Adequate levels of nutrients are essential for a healthy ecosystem and to support productive sustainable fisheries, but excess nutrients can cause hypoxia, which is a decrease in the levels of oxygen available in the water. Hypoxia can adversely impact fish and other aquatic life, and creates what is commonly called the "dead zone" in the Gulf. An oversupply of nutrients can also increase the amount of algae present and result in HABs. These algal blooms, some of which are called "red tides," can kill fish, cause skin rashes in swimmers, result in respiratory problems for beachgoers and render fish and shellfish unsafe to eat. When these blooms of algae die, they sink to the bottom and decompose—a process that consumes more oxygen. Reducing the excess nutrient levels in inland waters, before they reach the Gulf, is key to reducing the size of the "dead zone," decreasing the extent and frequency of HABs, and maintaining the health of coastal waters.

Important water quality priorities for the state include:

- Conduct regional nutrient characterization studies to identify and evaluate strategies to manage nutrient levels.
- Establish a Gulf states approach to develop coastal nutrient criteria and management.
- Support the goals and actions of the Mississippi/Gulf of Mexico Watershed Nutrient task force as identified in the Gulf Hypoxia Action Plan and promote the exchange of information and technology between the upper and lower Mississippi River Basin states and organizations.
- Develop a long-term "watershed to Gulf" monitoring program.
- Enhance and expand offshore monitoring/observing system.
- Maintain safe swimming beaches.
- Learn more about waterborne, disease-causing microorganisms (pathogens) and their sources. Then develop and implement proactive strategies to reduce people's exposure.
- Reduce the effects of HABs by improving the state's ability to detect, track, forecast and mitigate their movement and effects.
- Continue to monitor the conditions of the Gulf and the safety of seafood, with testing protocols and results being clearly and continuously communicated to the public.

Replenish and Protect Living Coastal and Marine Resources

Measures taken must include efforts to protect and restore the Gulf Coast's ecosystems from both natural and human-made disasters. It is imperative that the knowledge, skills, energy, passion and relationships of individuals and groups at all levels be leveraged for the successful creation and implementation of the strategies to keep the Gulf Coast viable and continue its legacy to the state and the nation.

Key priorities for protecting and replenishing coastal and marine resources in the state include:

- Protect and restore important habitats, including wetlands, seagrass beds and oyster reefs.
- Develop a strategic coastal and marine spatial plan that protects critical habitats and life stages.
- Enhance and improve existing long-term monitoring programs.

Enhance Community Resilience

The Mississippi Gulf Coast is an interdependent web of the relationships among the people, the economy and the environment. Mississippi's citizens and officials have long debated, planned and implemented ways to make the Gulf Coast more environmentally healthy and economically vibrant. The Gulf Coast people have the ideas and energy for the restoration and sustainability of their coastal ecosystems. It is imperative that the knowledge, skills, energy, passion and relationships of individuals and groups at the local level be leveraged to create and implement strategies to build community resilience. The power of local input is a common thread among the various plans, strategies and reports prepared before and since the *Deepwater Horizon* incident. Moving forward, local implementation is critical to the success of any restoration effort in the Gulf.

The state's priority actions for enhancing community resilience include:

- Implement the remainder of MsCIP, including voluntary relocation.
- Support GOMA Resilience Actions.
- Promote development of working waterfronts consistent with NOAA's Smart Growth Initiative.
- Establish and enhance capacity-building programs for local governments.
- Promote and enable locally driven solutions.
- Enhance communication of risk and science information to promote resilience.
- Identify and support critical research initiatives.
- Provide uniform storm surge and wave evaluations for all Gulf Coast communities.
- Develop a Gulf-wide sediment budget.
- Identify actions that would help provide sustainable reductions in storm surge risks for high-risk populations.

Texas

Background

The Texas coastal zone includes 367 miles of Gulf shoreline and more than 3,300 miles of bay, estuary and lagoon shorelines. It stretches from Port Arthur, near the Louisiana state line, to Brownsville on the banks of the Rio Grande River. The 18 counties within the Texas coastal zone make up only one-tenth of the total land area of the state, yet are home to approximately six million residents, which represent roughly 26 percent of the state's population of 25 million.[220] The coastal zone includes a complex system of barrier islands and peninsulas, with Padre Island being the longest undeveloped barrier island in the world. Texas also has more than 191,000 miles of rivers and streams, seven major estuaries, and about 200 major springs. Of the 12 major watersheds in Texas, only one does not flow to the Texas coast.

The Texas coast is one of the most productive and ecologically distinctive shorelines in the world. It contains 12 distinct eco-regions, covering 268,500 square miles.[221] It is richly endowed with natural resources, including sand dunes, vast wetlands and aquatic habitats. With nearly two-thirds of the state's Gulf shoreline protected in parks, wildlife refuges and natural areas off-limits to development,[222] the Texas coast is a natural wonder. It is home to blue crabs, oysters, pelicans, plovers, shrimp, the rare whooping crane and the Kemp's Ridley sea turtle, which nests only on western Gulf beaches. Much of the Texas coast is composed of important habitats for breeding, wintering and migrating birds. Coastal Texas prairie wetlands, coastal marshes, bays and seagrass beds are the second most frequented wintering area for waterfowl along the Gulf of Mexico.

The natural beach/dune system of barrier islands and peninsulas on the Texas coast is the first line of defense against coastal storms—such as Hurricane Ike, which caused over $27 billion of damages, making it the second costliest U.S. hurricane between 2004 and 2010.[223] Vegetated and unvegetated sand dunes provide protective barriers for adjacent land and inland water against the erosive action of waves, winds and storm surges.[224]

Dunes along the Texas coast also serve as vital habitat for numerous native plants and species of migratory and shore birds, and provide nesting areas for several species of endangered sea turtles. Wetland and aquatic habitats are also essential components of the Texas estuarine and inland systems. These habitats include salt, brackish, intermediate and fresh marshes, estuarine and palustrine scrub shrubs, bottomland hardwoods, sand and mudflats, mangroves, and seagrasses. Texas wetlands serve as nursery ground for over 95 percent of the recreational and commercial fish species found in the Gulf of Mexico. They provide breeding, nesting and feeding grounds for more than a third of all threatened and endangered animal species. They also provide permanent and seasonal habitat for a variety of wildlife, including 75 percent of North American bird species.[225]

Roughly one-third (423 miles) of the GIWW is also located in Texas. The GIWW is a navigable inland waterway that stretches 1,300 miles along the Gulf Coast from Florida to Texas.[228] The ecosystems adjacent to the GIWW contain sand dunes, wetlands, seagrasses, coastal prairies, oak mottes, sand and mud flats, and bay-estuary-lagoons that provide habitat for many wildlife and fish species.

THE COASTAL REGIONS OF TEXAS

The Texas coast lies within three very distinct coastal regions, which differ in both climate and geology:

- The humid *Upper Coast* extends from the Sabine River at the Louisiana state line to the vicinity of the mouths of the Brazos and Colorado Rivers. It is generally a sediment-starved system that has some of the highest erosion rates in the nation—up to 30 feet per year in some areas.[226] Between the 1950s and 1989, subsidence and sea-level rise in the Upper Coast converted 26,400 acres of wetlands in the Galveston Bay system to open water and barren flats.
- The less humid *Middle Coast* extends from the mouths of the Brazos and Colorado Rivers southward to the north shore of Baffin Bay. Erosion rates are lower than along the Upper Coast due to more sediment in the system; however, wetland loss remains an important concern.
- The semi-arid *Lower Coast* extends from the southern shore of Baffin Bay south to the Rio Grande delta area. While it generally has a sufficient supply of sediment, it lacks the dune vegetation to hold the sand dunes in place. The Lower Coast also contains the Laguna Madre of Texas, the only hyper-saline coastal lagoon on the North American continent. The lagoon is known for its vast seagrass meadows, huge wintering bird populations and bountiful fishing grounds.[227]

The GIWW is vital to the economies of Texas and the nation as a whole. It is the nation's third busiest waterway and an important component of the diversified Texas transportation system. Thirteen percent of the nation's freight traveled the Texas GIWW in 2006, consisting of approximately 74 million tons of cargo with an estimated value of $25 billion.[229] Two-thirds of the nation's petrochemicals are produced by plants along the Texas portion of the GIWW, and the waterway handles 90 percent of all gasoline shipped to the Lower Rio Grande Valley.[230]

The Texas coast is also home to four of the top 10 ports in the country (based on total cargo tonnage),[231] including the ports of Houston, Corpus Christi, Beaumont and Texas City.[232] Texas ports generate over $9 billion in federal tax revenue annually. The four busiest Texas ports handled almost 400 million tons of cargo in 2009.[233] Galveston Bay supports one of the largest metropolitan areas in the United States[234] as well as the Port of Houston, which is the second largest port in the nation.[235]

The Port of Houston handles more total foreign trade and imports than any other U.S. port.[236] It accommodated 220 million tons of cargo and 7,700 vessels in 2009.[237] It provides more than 287,000 jobs and $30 billion to the economy annually. The area is also home to the nation's largest concentration of oil refineries[238] as well as a chemical industry that is ranked first in the nation in size and production.[239]

Ports and petrochemical production are only a part of the economic benefit derived from the Texas coast. Texas commercial fishing fleets bring in more than $150 million of fish and shellfish annually.[240] Eighty-two percent of the shrimp in the United States come from the Gulf states,[241] with Texas supplying 89.7 million pounds per year,[242] or almost 30 percent of the U.S. total shrimp landings. The recreational fishing industry is another important part of the Texas coastal economy with saltwater sport fishing generating almost $2 billion annually.[243] The number of annual saltwater fishing permits increased over 7.1 percent between 2006 and 2010.[244]

Oyster reefs in the bay systems along the Texas coast have both ecological and commercial importance. They provide valuable ecological services, such as supplying habitat for other commercial and recreationally important finfish and shellfish species, improving water quality, reducing turbidity and providing shoreline protection (intertidal reefs) from storm-induced erosion. The annual oyster harvest is approximately 5.7 million pounds of meat worth over $19 million. Prior to Hurricane Ike in 2008, Galveston Bay was responsible for producing approximately 85 percent of the Texas commercial oysters. Approximately 50 percent of the consolidated oyster reefs (8,000 acres) in Galveston Bay were lost due to hurricane-induced sedimentation, resulting in a 46 percent reduction in commercial oyster landings in the season following the hurricane's landfall.

Tourists visiting the Texas coast spend more than $7.5 billion annually for beach recreation, bird watching, fishing and eco-tourism.[245] The coast accounts for more than one-quarter of the total travel expenditures in Texas, making it the second most popular tourist destination in the state.[246] In 2009, Texas Gulf Coast tourism accounted for more than 27 percent of the state total of hotel rooms. It has consistently accounted for around 28 percent of the annual direct travel spending in Texas,[247] which amounts to approximately $14.5 billion[248] annually. Since the opening of the Galveston's first cruise line terminal in 2000, more than a million people have sailed from the Port of Galveston on cruise ships,[249] which contributed more than $1.05 billion to the Texas economy in 2009.[250]

State Priorities

The Task Force Strategy lays out four overarching goals. Texas's priority actions for each goal are described below.

Restore and Conserve Habitat

Erosion is one of the most critical issues facing the Texas coast. Sixty-four percent of the total Texas coast is eroding at an average rate of 5.9 feet per year, with some areas experiencing losses greater than 30 feet per year.[251] One of the primary contributors to this erosion is a lack of sediment in the coastal zone and shoreline currents to balance the effects of coastal storms and sea-level rise.

Erosion decreases the natural resiliency of the Texas coast to weather and storm events; limits public beach access; impacts tourism; threatens roads, structures and property; and affects future growth and development of coastal communities. Erosion has already consumed 26 miles of State Highway 87, which once connected Galveston to Sabine Pass.

Wetland loss is another major threat to Texas habitats. An estimated 4.1 million acres of wetlands existed in Texas in the 1950s. Less than 3.3 million acres were present by the early 1990s.[252] During this period, estuarine wetlands declined from 165,000 acres to 130,400 acres along the Texas coast.[253] Causes of wetland loss include sea-level rise, erosion, saltwater intrusion, urban and rural development, nonpoint source pollution, invasive species and agriculture.

The coastal prairie ecosystem in Texas is listed as critically imperiled by major conservation organizations. Approximately 6.5 million acres of coastal prairie existed in

Texas before settlement occurred.[254] Currently less than 1 percent of the coastal prairie remains due to land conversion for cattle grazing or growing crops.[255]

The state's priorities for restoring and protecting habitat include:

- Protect and strengthen existing dunes (increase height and stability) and create new dunes.
- Restore wetlands, barrier islands/peninsulas/headlands and beaches through the beneficial use of dredged material.
- Restore sediments to river delta wetlands.
- Enhance Gulf beach monitoring and maintenance.
- Implement conservation projects to protect coastal habitats, especially on the Middle and Lower Coasts.
- Implement watershed-scale restoration projects that restore freshwater inflows and further restrict saltwater intrusion.
- Restore, protect and enhance severely depleted oyster reefs in the Texas coastal ecosystems because of their ecological and commercial significance.
- Maintain natural Gulf and bay shoreline areas for public recreation while increasing access and opportunities.
- Construct protective breakwaters, especially along the entire 40-mile length of the GIWW along Bolivar Peninsula.
- Adopt and implement a plan for the rapid reporting and long-term management of invasive species.
- Develop public and private strategies that build broad-based support for successful and adaptive management, restoration, and conservation.
- Promote, develop, maintain, monitor and enhance natural habitats along the Texas Coast through restoration and acquisition.

Restore Water Quality

Besides being a source of beauty and wonder, Texas waters are an essential life-supporting resource for animals, plants and humans. Healthy aquatic ecosystems depend on careful and effective water management. The population of Texas is expanding rapidly, bringing incredible pressure to bear on all of the state's natural resources, especially water, which intensifies competition for water among industry, urban areas, agriculture, riverine and coastal wetland wildlife, and fisheries. Protecting water quality and environmental flow regimes will help minimize adverse impacts to aquatic ecosystems. Resource managers should partner with stakeholders to develop science-based strategies designed to protect the aquatic ecosystems of Texas.

The state's priority actions to address water quality issues include:

- Work with public and private entities to integrate planning and management of groundwater, spring, stream, wetland, estuarine and marine ecosystems.
- Protect, maintain or restore appropriate watershed and hydrologic conditions to support healthy aquatic ecosystems.

- Establish and maintain cooperative strategies to incorporate long-term plant, fish and wildlife needs in all statewide, regional and local watershed planning, management, and permitting processes.
- Ensure that the Texas Surface Water Quality Standards increasingly incorporate biological data to protect the health and productivity of Texas waters.
- Identify instream flow and freshwater inflow regimes, especially for estuarine systems, to supply water, sediment and nutrients needed to support coastal habitats under natural and modified scenarios and salinity levels necessary to maintain a sound ecological environment.
- Ensure environmental flow standards protect instream flow and freshwater inflow regimes that are adequate to support fish and wildlife resources.
- Focus restoration of water quality and quantity on watersheds associated with bay ecosystems, including Galveston Bay, San Antonio Bay, Corpus Christi/Nueces Bay, Matagorda (East and West) Bay, Laguna Madre (Upper and Lower) and Aransas Bay.
- Collaborate with Louisiana and Mexico on restoration activities for the Rio Grande River and Sabine Lake.
- Increase implementation of BMPs and native buffers in watersheds.
- Develop and implement watershed-wide nutrient and contaminant reduction strategies. Evaluate and encourage agricultural, residential, industrial and commercial BMPs to reduce nutrient and contaminant inputs to the Gulf.
- Reduce impacts of HABs by improving detection, tracking and forecasting ability.
- Quantify the areal extent of land inundated during flood events to determine the benefits and impacts.
- Continue and enhance beach water quality monitoring.

Replenish and Protect Living Coastal and Marine Resources

The Texas coast has tremendous biodiversity. More than 457 species of fish and 343 species of invertebrates are found in the estuarine and marine waters of Texas. Texas currently operates 93 state parks and natural areas, 51 wildlife management areas,[256] and eight fish hatcheries.[257] These cover over 1.4 million acres that are managed in the public trust for recreation and conservation. As the population of Texas increases, so do land fragmentation, loss of open space, genetic isolation, habitat degradation and other impacts. Protecting and enhancing the living coastal and marine resources is critical to Texans.

The state's priorities to help replenish and protect its coastal and marine resources include:

- Continue to enhance and expand long-term monitoring (>35 years) of dependent and independent fisheries data within Texas bays and Texas Territorial Seas.
- Standardize monitoring (for success) of marsh/wetland and other restoration projects.
- Expand submerged habitat mapping and monitoring by expanding current side-scan sonar mapping efforts through all Texas bay systems and Texas Territorial Seas.

- Expand oyster habitat monitoring to include biomass components, disease, community parameters and commercial harvest potential.
- Quantify and develop indicators of ecosystem health (offshore, near-shore and estuarine) to use in adaptive management.
- Maintain and enhance wildlife and fisheries stocks in estuarine and marine habitats (bays, estuaries, natural and artificial reefs and the Gulf of Mexico).
- Acquire additional wildlife management areas for habitat conservation, demonstration and public hunting, and secure related water rights.
- Protect and assist in the recovery of threatened, endangered and high-priority species (e.g., marine sea turtles, terrapins, and other plant, fish and wildlife species) and their habitats.
- Develop and incorporate a coast-wide habitat interface into a routine monitoring program to identify potential impending large-scale losses of habitat and associated biological communities.
- Create an active network/database for ongoing monitoring, restoration and research efforts in Texas to maximize potential leveraging of effort and funds.
- Publish, disseminate, and promote guidelines and protocols for habitat restoration and management.
- Concentrate on-the-ground conservation efforts on landscapes of high biological value, such as watersheds, recharge zones, wildlife corridors and migratory flyways.
- Support conservation actions that mitigate anticipated climate change impacts to plants, fish and wildlife.
- Encourage development of renewable energy projects that do not adversely affect plant, fish and wildlife communities.
- Assess contaminants, bioaccumulation and diseases in fish and shellfish.
- Promote, develop, monitor and enhance the artificial reef potential of Texas.

Enhance Community Resilience

Population levels within the Texas coastal zone are expected to increase by an additional six million over the next 30 years.[258] The coastal region's largest urban centers—such as Houston and Corpus Christi—are continuing to grow, and the expected population increase will place greater demands on the area's economic and environmental resources.

Some of the most destructive hurricanes in U.S. history since 1875 have struck the Texas coast. The densely populated upper Texas coast is the most vulnerable area. Based on historical records dating back to 1871, on average this area is affected by a tropical storm or hurricane every 2.62 years. Since 1851, the upper Texas coast has received 56 percent of all tropical cyclone landfalls.[259]

However, none of the Texas coast is immune to hurricanes. Since 1953, Texas has experienced 23 federal disaster declarations due to hurricane and tropical storm events. The destructive impacts from hurricanes are not confined to the coastal zone either. As a hurricane or tropical storm moves inland and weakens, it can drop heavy amounts of rain, causing extreme flooding and extensive damage to the central areas of the state.

Another major threat to the Texas coast is subsidence, particularly in the Galveston-Houston vicinity, where the area has experienced a loss of elevation as much as 10 feet from 1906 to 2000. Subsidence increases coastal communities' risk of inundation and saltwater intrusion from storm surge. It also creates and exacerbates shoreline erosion. In Galveston, local sea levels are rising at approximately 0.02 feet per year.[260] At this rate, sea levels in the area could be approximately 1.75 feet higher by the year 2100.

EDUCATING COASTAL DECISION-MAKERS

Educating coastal decision-makers on community resiliency and how it depends on ecosystem resiliency can be one of the most important steps in planning and being proactive. Through participation on the GOMA Coastal Community Resilience Workgroup, Texas has helped develop several tools for coastal decision-makers.

The Texas Storm Smart Coast Network provides information to coastal communities on preparing and recovering from storm events. A Community Resiliency Index is available for communities to conduct a self-assessment to develop an idea of how well they might reach and maintain an acceptable level of functioning after a disaster or from potential impacts resulting from sea-level rise.

Priority actions for enhancing community resilience in Texas include:

- Complete the development of a coastal and marine spatial plan to address the unique needs of Texas citizens in the planning process.
- Increase public awareness of the potential risks of living in vulnerable coastal areas, how to prepare for those risks and how to interpret and adapt to future risks.
- Complete development of the *Texas Homeowners Handbook to Prepare for Coastal Natural Hazards*, which will provide information on protecting homes and property from coastal hazards.
- Continue the development of educational workshops for the public so they can learn more about the risks of living in a coastal environment and adapt to future risks.
- Assist coastal communities in developing plans to prevent, adapt to and rebound from disasters, negative economic and social/cultural changes, and chronic long-term ecological stressors.
- Develop post-disaster redevelopment plans that can help guide communities to become more resilient as they make decisions that affect long-term recovery and redevelopment.
- Continue developing a capacity-building program to support ecosystem resilience by bringing together local planners, emergency managers, floodplain managers, building code officials and others.
- Coordinate and integrate coastal regional planning committees and provide technical support and tools to better educate communities about the natural and beneficial use of their coastal ecosystems.

APPENDIX C. SCIENCE TO SUPPORT GULF OF MEXICO ECOSYSTEM RESTORATION

Introduction

The goals highlighted within the Strategy are oriented around the many components of the ecosystem, including the human component. Fundamental to the success of the Strategy is ensuring that it has a robust and defensible scientific foundation. Given the interconnected nature of the Gulf ecosystem, issues that relate to one goal (e.g., coastal habitats) often have direct bearing on other goals (e.g., living marine resources). The scientific activities highlighted in this appendix—monitoring, modeling and research—overlap among many of the goals and will provide the knowledge and understanding needed to make and implement informed decisions.

Science Priorities

A long-term Gulf of Mexico monitoring program will support a variety of restoration and protection project alternatives and support valid predictions to protect human life and restore the ecosystem. Such a program should be used to determine baseline conditions for inland watersheds and estuarine, coastal, and offshore waters and be used to measure change, assess project effectiveness and support adaptive management decisions for Gulf restoration. A Gulf of Mexico modeling network should also be developed that increases certainty in forecasts and estimates of ecosystem services at a variety of stages along the restoration continuum for decision-makers and the public.

Further, research and basic discovery is needed to improve understanding of the ecosystems in the Gulf and how they can be sustained when the Gulf is undergoing extreme adverse conditions, including human and natural disasters (such as oil spills or hurricanes) and climate change. Focused research on human impacts, solutions, and risk is needed, as well as information about the economic impacts to humans and ecosystem services. There should be a strong reliance on basic research on such subjects as ecosystem loss, adaptability, variability and resiliency in all forms.

These activities will promote learning and help guide the planning, implementation and evaluation of the restoration and protection efforts articulated in the goals of this Strategy, as well as future restoration efforts in the Gulf. As the monitoring, modeling and research priorities identified below are implemented, shared learning among all stakeholders should occur over multiple iterations of the adaptive management process. Science is critical in the development of the projects, but equally critical is to determine the scientific and cost effectiveness of restoration projects and not repeat mistakes or ineffective efforts. The priorities outlined here highlight preliminary needs supporting Gulf ecosystem restoration. Addressing these needs should include an assessment of the existing capacity (monitoring assets and data streams, model inventories, and research results) and build upon this capacity. Project-specific elements will be defined as restoration and conservation projects are developed.

Monitoring

Performance indicators are used to determine both system-wide and project-level monitoring needs. Monitoring is conducted to address decision-critical uncertainties and to parameterize models needed to assess performance. The data acquisition needed to support monitoring and modeling typically exceeds the resources available to meet those needs and, thus, requires prioritization. The following were identified as high priorities:

Monitoring Programs

- Collect information about existing watershed, estuarine, coastal, offshore, and habitat monitoring programs across the Gulf (e.g., the Gulf Coast Ocean Observing System, the Coastwide Reference Monitoring System) and identify gaps that should be filled to better support adaptive management.
- Recommend ways to integrate these programs and fill gaps to establish a comprehensive network that can provide the information necessary for managers to make informed decisions, adapt their actions as needed, and facilitate effective stewardship of Gulf ecosystem resources. Identify gaps in the monitoring programs that need to be filled to support adaptive management.
- Use a hypothesis-based approach for assessment of system performance.
- Foster data comparability, consistency, and standardization across programs, projects, and habitats.
- Improve data dissemination and visualization tools to provide information to resource managers.

Monitoring Variables

- Collect high-resolution topographic, bathymetric, geodetic and tidal data to develop and maintain (with frequent updates over time) high-quality digital elevation models for the Gulf of Mexico that reflect and quantify changes to a dynamic (i.e., constantly changing) land and seafloor.
- Collect water, sediment, pollution and nutrient loading data from a comprehensive network of inland stream gauges and wetland stations, as well as nearshore/offshore ocean observing stations that also acquire wave, current and sediment transport characteristics.
- Monitor the capacity of Gulf Coast communities to plan and implement resilience programs, to communicate with the public, and to provide feedback to decision-makers and resource managers.

The task force identified specific data needs by theme. These are identified in Table 1. Many of these variables can serve multiple goals and would be considered higher-priority; however, each restoration project should be assessed to determine that it incorporates the monitoring elements required to determine project efficacy.

Modeling

As highlighted throughout the Strategy, models can be used to modify or adjust restoration and protections actions, and to inform analysis and guidelines to the efficacy of different restoration strategies/projects (such as re-establishment or modification of freshwater flow, nutrient loads, suspended sediment deposition, storm buffers, barrier island restorations). Modeling is used to understand system processes, make predictions related to different management/restoration scenarios/projects, and guide monitoring. Modeling can also be used to address future uncertainties, like the effects of relative sea-level rise. As with the use of adaptive management overall, model assumptions and uncertainties should be clearly articulated to ensure that planning and decisions are fully informed. The following were identified as high priorities:

Table 1. Specific Data Acquisition Needs

	Habitats (coastal)	**Habitats (inland) And watersheds**	**Living coastal and marine resources and offshore environments**	**Coastal communities (including storm buffers)**
Physical				
Sediment, nutrient, pollutant loads, and freshwater flow rates	x	x	x	x
Land:water ratios	x	x	x	x
Topography/bathymetry	x	x	x	x
Shoreline position and form and dimensions of beaches and dunes and barrier islands	x		x	x
Erosion and accretion rates	x			x
Seafloor change	x	x		
Hydrology (water surface elevation, current velocity, wave characteristics, salinity, temperature)	x	x	x	x
Meteorology	x		x	
Air quality		x	x	
Marsh elevation (accretion, subsidence, sediment elevation table)	x		x	x
Relative sea-level rise rates (subsidence and global sea-level rise)	x	x	x	x
Geodetic vertical datum	x	x		x
Biological				
Invasive species	x		x	
Fisheries composition/abundance/ diversity/ productivity/tissue contaminants	x		x	
Fisheries landings			x	x
Wildlife and living marine resources abundance/diversity and distribution (including sentinel species)	x	x	x	
Plant community composition/abundance/diversity/productivity	x	x	x	

	Habitats (coastal)	Habitats (inland) And watersheds	Living coastal and marine resources and offshore environments	Coastal communities (including storm buffers)
Benthic macroinvertebrates or key benthic assemblages	x		x	
Phytoplankton, harmful algae species occurrence, toxin production	x	x	x	
Zooplankton	x		x	
Pathogens	x	x	x	
Microbial ecology		x	x	
Chemical				
Water quality (nutrients, ammonia, silica, turbidty, total suspended solids, water clarity, contaminants [e.g. PAHs, PCBs], metals, dissolved oxygen, salinity, temperature, depth, conductivity, secchi depth, photosynthetically active radiation, pH, chlorophyll a, carbon)	x	x	x	
Coastal, nearshore and offshore seafloor sediment characteristics (sediment composition, bulk density, organic matter, total carbon, total nitrogen, phosphorous, grain size, total organic carbon, sediment toxicity)	x	x	x	
Habitat				
Habitat Classification (including classification of impaired habitats)	x	x	x	
Areal extent of essential habitat	x	x	x	
Areal extent of sustainable land use		x		
Socioeconomic				
Socioeconomic data on habitat and living marine resources			x	x
Social and community capacity for emergency preparedness				x
Population and development in high-risk or hazardous areas				x
Community networking capacity				x
Environmental awareness and attitudes, as well as barriers to adopting resilience practices				x

Modeling Programs

- Document existing watershed (surface water), groundwater, estuarine, offshore, erosion and habitat models across the Gulf and encourage collaboration.
- Use models to modify or adjust restoration and protection actions, and to inform analysis and guidelines for the efficiency of different restoration strategies/projects (such as re-establishment of freshwater flow, nutrient loads, suspended sediment deposition, storm buffers and barrier island restorations) in an adaptive management framework.

- Promote fully coupled surface water–groundwater models linked to watershed, coastal, biological, ecological and offshore models to support adaptive management strategies and evaluate the effects of restoration projects on the ecosystem over time.

Modeling Input

A comprehensive monitoring network with organized data management and quality assurance/control can provide the necessary input for models. The models can also be used to guide data collection and monitoring programs by evaluating the reduction in predictive uncertainty by the inclusion/exclusion of existing data and proposed monitoring sites (i.e., the "worth" of the data).

Modeling Needs

- *Predictions and adaptive management*
    - Employ ecosystem modeling to support planning and explore relationships between management actions and resource response.
    - Develop models and other decision support tools to predict the amount of water, sediment and nutrients needed by coastal habitats to support wetland and marine organisms:
        - Under realistic sea-level rise scenarios
        - Including riverine and marine sources
        - Incorporating water quality data and hydrologic flow data
    - Develop models to predict ecosystem resilience under different stressor paradigms, including:
        - Climate change and sea-level rise
        - Subsidence
        - Storm intensity and frequency, associated wave action, and rainfall
        - River discharge and associated sediment, nutrient and pollutant loading
    - Test models with experimental, natural and hypothetical disturbance events.
    - Identify and address critical model limitations and uncertainties, including compounding uncertainties when linking with one or more models and/or into future years.
    - Develop uniform methodologies for including relative sea-level rise considerations into modeling and project planning for sustainable storm buffers including wetland accretion.
    - Improve models for predicting coastal response to relative sea-level rise and storm impacts.

- *Physical and biological models*
    - Develop/enhance storm surge, wave and coastal erosion models that can be used in developing risk assessments of hurricane, storm surge and wave impacts.
        - Couple storm surge and wave modeling for hindcasting (used in Katrina), and for flood modeling for risk assessments
        - Develop a coherent and robust model for the entire Gulf (all states)
        - Include topographic data

  - Include bathymetric data
  - Consider offshore topographic data
    - Offshore wave models (e.g., WAM)
    - Storm surge models (e.g., ADCIRC)
    - Nearshore wave models (e.g., STWAVE or UnSWAN)
- Develop and enhance hurricane wind and pressure field models.
- Evaluate the sustainability of storm buffers and barrier islands as habitats using coastal erosion and morphologic evolution models.
- Focus a global climate model to address Gulf needs.
- Develop models to understand the hydrologic regime of targeted watersheds of the Gulf of Mexico. Surface water–groundwater integrated models of coastal systems with transport included.
- Model impacts of hydrological restoration and protection (e.g., diversions, levee realignments) on diversity and/or production of living resources (e.g., vegetation, fish, shellfish).
- Model impacts of habitat loss (e.g., marsh degradation) on diversity and/or production of living resources (e.g., vegetation, fish, shellfish).

Research

The influences of ecosystem variability, gaps in knowledge, and inadequate understanding of complex ecosystem functions and responses cause uncertainty that can greatly influence risk in management actions, thereby increasing the need for ecosystem research to support management decision-making.

Research Programs

Testing underlying assumptions of ecosystem behavior is an integral component of supporting research. Numerous hypotheses have been identified from previous studies conducted across the Gulf Coast; however, results should be focused on clearly meeting the Strategy needs. Supporting research should be directed at reducing scientific uncertainty to improve confidence in modeling and monitoring tools and ultimately management actions. Additionally, a key weakness that basic research must help address is simply discovering what ecosystems exist in the Gulf that are or may be impacted. It is essential to integrate monitoring, modeling and research development activities, from the initial stages of restoration and protection planning, in order to support adaptive management decision-making.

Research Needs

Specific research needs underpinning restoration goals are described in Table 2. Addressing these needs would serve to support broader ecosystem-wide restoration efforts. Additional effort should be directed to addressing questions that inform discrete restoration projects. Research efforts should leverage ongoing planning efforts and programs, where appropriate (e.g., the Gulf of Mexico Research Plan, the Gulf Research Initiative).

Table 2. Research Needs to Support Restoration

Resilience	• **Develop** a shared vision of ecosystem resilience • **Identify** key determinants of resilience for estuarine, coastal wetland, forested ridges and barrier shoreline habitats • **Examine** the relationship between ecological and human community resilience **Examine** how land change, sediment types, anthropogenic modifications, and flood and storm damage risks can affect ecosystem resilience • **Quantify** (spatially and temporally) relative sea-level rise (including subsidence) rates • **Determine** the relationship between shallow stratigraphy and natural processes • **Quantify** the Gulf of Mexico sediment budget • **Quantify** the amount of sediment and nutrients that bypass wetlands and are discharged offshore • **Identify** those nutrient levels that are excessive and lead to negative impacts in coastal wetlands, seagrasses and fisheries, and contribute to harmful algal blooms and hypoxic conditions • **Establish** the key relationships between nutrients, sediment and salinity as they relate to water clarity, optimal ecological function, optimal distribution of habitats and species • **Provide** a more comprehensive understanding of life histories of affected living marine resources, food web dynamics, and essential habitat conditions, and what habitat conditions are essential • **Investigate** surge/wave /vegetation interactions and the influence on geomorphologic evolution of landforms • **Identify** agricultural practices that use less fertilizer, water and pesticides and preserve topsoil
Risk	• **Develop** a better understanding of critical landscape and geologic features (i.e., geomorphic, geologic, biological, physiochemical, engineered) to reduce storm risk • **Develop** a better understanding of engineering tools used in storm risk assessment such as storm surge models and coastal erosion models • **Improve** understanding of the impact of physical development on flood outcomes • **Understand** vulnerability of communities to storm surge, land loss, subsidence and sea-level rise • **Refine** risk or vulnerability indices **Examine** approaches to communicate to coastal residents and decision-makers what puts them at risk and what they can do to reduce risk, and identify constraints and incentives to pursue resilient behaviors • **Identify** features that make shoreline habitats more or less vulnerable to stressors such as sea-levelrise, high wave energy, storm surge, coastal erosion and sediment loss • **Establish** the relationship between ecosystem restoration and community storm risk reduction • **Identify** the cultural, economic and social impacts of relocation of people out of risky coastal areas

Ecosystem Services	• **Determine** processes and functions supported by Gulf coast habitats and the degree to which optimal function and provision of priority ecosystem services is presently occurring • **Measure** rates and processes that reflect wetland ecosystem condition and the ecosystem services they provide, and consider functional equivalence • **Determine** assessed value of fishing, recreation and ecosystem services that are provided to the community • **Determine** the relationship between nutrient loading and ecological function, along with the potential for: • **Development** of hypoxia and associated impacts on the benthos • **Development** of harmful algal blooms • **Loss** of seagrass meadow acreage • **Change** in fisheries productivity • **Change** in soil composition • **Develop** ecological indicators for ecosystem structure and function • **Identify** research-based criteria for meeting water quality standards • **Identify** most efficient paths for various community types to improve resilience
Restoration and Hydro-logic Modifi-cation	• **Examine** impacts of upstream hydrological modification and varying freshwater flow on estuarine vitality • **Identify** optimal water timing, quality and quantity to support sustainable ecosystem habitats • **Determine** the relationship between varying scales of river diversion and ecological function and resilience of wetlands • **Examine** the function and resilience of emergent wetlands and barrier shorelines that have been restored by sediment augmentation **Examine** how upstream reservoir and dam management practices impact delivery of sediment andfreshwater to coastal ecosystems • **Identify** the storm buffering consequences of common coastal engineering projects • **Identify** the optimal size of natural buffers for water filtration **Examine** the ecological function and resilience of other habitat restoration efforts such as oyster reefs, coral reefs, vegetative plantings and submerged aquatic vegetation
Climate	• **Develop** uniform methodologies for including relative sea-level rise considerations into modeling and project planning • **Develop** uniform methodologies for including climate change-induced variations in precipitation, evapotranspiration, and changes in storm intensity and frequency into future planning decisions

List of Abbreviations

Acronym	Definition
ADCIRC	Coastal Circulation and Storm Surge Model
BMP	best management practice
CEAP	Conservation Effects Assessment Program
CERP	Comprehensive Everglades Restoration Plan

Acronym	Definition
CIAP	Coastal Impact Assistance Program
CNMP	Comprehensive Nutrient Management Plan
CRE	Climate Ready Estuaries
CWPPRA	Coastal Wetlands Planning, Protection and Restoration Act
DOI	United States Department of the Interior
DOT	United States Department of Transportation
ECA	Emissions Control Area
EPA	United States Environmental Protection Agency
ESA	Endangered Species Act
GIWW	Gulf Intracoastal Waterway
GOMA	Gulf of Mexico Alliance
HAB	harmful algal bloom
HUD	United States Department of Housing and Urban Development
IEA	integrated ecosystem assessment
LACPR	Louisiana Coastal Protection and Restoration
LGAC	Local Government Advisory Committee
MRBI	Mississippi River Basin Initiative
MsCIP	Mississippi Coastal Improvement Program
NASA	National Aeronautics and Space Administration
NEJAC	National Environmental Justice Advisory Council
NEP	National Estuary Program
NGO	nongovernmental organization
NOAA	National Oceanic and Atmospheric Administration
NRDA	Natural Resource Damage Assessment
PAH	polycyclic aromatic hydrocarbon
PCB	polychlorinated biphenyl
RCSD	Regional Climate Services Director
SEAMAP	Southeast Area Monitoring and Assessment
STWAVE	STeady-state spectral WAVE
TMDL	Total Maximum Daily Load
UnSWAN	unstructured SWAN (Simulating WAves Nearshore)
USDA	United States Department of Agriculture
USFWS	United States Fish and Wildlife Service
USGS	United States Geological Survey
WAM	Wave Action Model

REFERENCES

[1] National Oceanic and Atmospheric Administration. (2011). *The Gulf of Mexico at a glance: A second glance.* <http://stateofthecoast.noaa.gov/NOAAs_Gulf_of_Mexico_at_a_Glance_report.pdf> p. 13.

[2] U.S. Energy Information Administration. (2010). n.d. Gulf of Mexico fact sheet. <http://www.eia.doe.gov/special/gulf_of_mexico/index.cfm> (As cited in Mabus, R.

America's Gulf Coast: A long term recovery plan after the Deepwater Horizon oil spill.<http://www.epa.gov/indian /pdf/mabus-report.pdf>)

[3] National Marine Fisheries Service. (2010). *Annual commercial landings statistics.* Years queried: 2007-2009. <http://www.st.nmfs.noaa.gov/st1/commercial/landings/ annual_landings.html> (As cited in Mabus, R. 2010. America's Gulf Coast: A long term recovery plan after the Deepwater Horizon oil spill. <http://www.epa.gov/indian/ pdf/mabus-report.pdf>)

[4] U.S. Army Corps of Engineers. (2010). *The U.S. waterway system: Transportation facts and information.* Navigation Data Center. <http://www.ndc.iwr.usace.army.mil/fact card/temp/factcard10.pdf> (As cited in National Oceanic and Atmospheric Administration. 2011. The Gulf of Mexico at a glance: A second glance. <http:// stateofthecoast.noaa.gov/NOAAs_Gulf_of_Mexico_at_a_Glance_ report. pdf>)

[5] National Oceanic and Atmospheric Administration. (2011). *The Gulf of Mexico at a glance: A second glance.* <http://stateofthecoast.noaa.gov/NOAAs_Gulf_of_Mexico_ at_a_Glance_report.pdf>

[6] National Oceanic and Atmospheric Administration. (2011). *The Gulf of Mexico at a glance: A second glance.* <http://stateofthecoast.noaa.gov/ NOAAs_Gulf_of_Mexico_ at_a_Glance_report.pdf> p.13.

[7] Barras, J., Beville, S., Britsch, D., Hartley, S., Hawes, S., Johnson, J., Kemp, P., Kinler, Q., Martucci, A., Porthouse, J., Reed, D., Roy, K., Sapkota, S. & Suhayda, J. (2003). *Historical and predicted coastal Louisiana land changes:* 1978-2050. USGS Open File Report. <http://cmbc.ucsd.edu/content/1/docs/LA_land_loss_trends_1956-20.pdf> (As cited in LACoast.gov. n.d. Mississippi River water quality: Implications for coastal restoration.<http://lacoast.gov/new/Data/Reports /ITS/MRWQ.pdf>)

[8] National Research Council of the National Academies. (2006). *Drawing Louisiana's new map: Addressing land loss in coastal Louisiana.* <http://www.nap.edu/catalog.php? record_id=11476>

[9] Couvillion, B. R., Barras, J. A., Steyer, G. D., Sleavin, W., Fischer, M., Beck, H., Trahan, N., Griffin, B. & Heckman, D. (2011). Land area change in coastal Louisiana from 1932 to 2010: U.S. Geological Survey Scientific Investigations Map 3164. <http://pubs.usgs.gov/sim/3164>

[10] U.S. Geological Survey and U.S. Environmental Protection Agency. (2011). *ESRI maps, National Hydrography Dataset, EPA analyses.* Courtesy of Stephen B. Hartley, USGS National Wetlands Research Center.

[11] National Oceanic and Atmospheric Administration. (2011). *Hypoxia in the Gulf of Mexico: Progress towards the completion of an integrated assessment.* Accessed September 23, 2011.<http://oceanservice.noaa.gov /products/pubs_hypox.html>

[12] Executive Order 13554. (2010). *Establishing the Gulf Coast Ecosystem Restoration Task Force.* <http://edocket.access.gpo.gov/2010/pdf/2010-25578.pdf>

[13] National Commission on the BP Deepwater Horizon Oil Spill and Offshore Drilling. (2011). *Deep water: The Gulf Oil disaster and the future of offshore drilling.* Report to the President.<http://www. gpoaccess.gov/deepwater/index.html>

[14] U.S. Bureau of Ocean Energy Management (Previously known as Minerals Management Service). (2011*). Outer Continental Shelf (OCS) Oil and Gas Production.* Accessed November 22, 2011.<http://www. boem.gov/BOEM-Newsroom/Offshore-Stats-andFacts/OCS-Oil-and-Gas-Production.aspx>

[15] National Marine Fisheries Service. (2010). *Annual commercial landings statistics.* Years queried: 2007-2009. Accessed October 4, 5, and 6, 2010. <http://www.st.nmfs.noaa.gov/st1/commercial/landings/annual_landings.html>

[16] Mabus, R. (2010). *America's Gulf Coast: A long term recovery plan after the Deepwater Horizon Oil Spill.* <http://www.epa.gov/indian/pdf/ mabus-report.pdf> p. 14.

[17] Mabus, R. (2010). *America's Gulf Coast: A long term recovery plan after the Deepwater Horizon Oil Spill.* <http://www.epa.gov/indian/pdf/ mabus-report.pdf>

[18] Oral history presentation, Bryant Celestine, Alabama-Coushatta Tribes of Texas, November 2011.

[19] Gulf of Mexico Alliance. (2008). Governors' action plan II for healthy and resilient coasts, 2009-2014. <http://www.gulfofmexicoalliance.org/ pdfs/ap2_final2.pdf>

[20] National Oceanic and Atmospheric Administration. (1975). *The coastline of the United States.* <http://shoreline.noaa.gov/_pdf/Coastline_of_the_US_1975.pdf> Includes length of general outline of seacoast; does not include freshwater coastlines.

[21] National Oceanic and Atmospheric Administration. (1975). *The coastline of the United States.* <http://shoreline.noaa.gov/_pdf/Coastline_of_the_US_1975.pdf> Includes length of outer coast shoreline, offshore islands, sounds, bays, rivers, and creeks to head of tidewater.

[22] U.S. Census Bureau. (2011). Census 2010. <http://factfinder2.census.gov/faces/nav/jsf/pages/index.xhtml>

[23] U.S. Census Bureau. (2011). Census 2010. <http://factfinder2.census.gov/faces/nav/jsf/pages/index.xhtml> Includes watershed counties that NOAA indicates have either (1) at least 15 percent of their land area within a coastal watershed, or (2) a portion of or an entire county accounts for at least 15 percent of a U.S. Geological Survey coastal cataloging unit.

[24] Mobile Bay Estuary Program. (2008). *State of Mobile Bay.* <http://www.mobilebaynep.com/images/uploads/library/State-of-Mobile-BayFinal. pdf>

[25] National Marine Fisheries Service. (2009). 2009—*fisheries economics of the U.S.* <http://www.st.nmfs.noaa.gov/st5/publication/fisheries_ economics_ 2009.html>

[26] National Marine Fisheries Service. (2009). 2009—fisheries economics of the U.S. <http://www.st.nmfs.noaa.gov/st5/publication/fisheries_ economics_2009.html>

[27] Mississippi-Alabama Sea Grant Consortium. n.d. MASGC focus on habitat restoration. <http://www.masgc.org/page.asp?id=219>

[28] Alabama Department of Tourism. (2010). Travel economic impact report2009. <http://www.alabama.travel/media/media_room/Report/ 2009TourismReport.pdf>

[29] U.S. Army Corps of Engineers. (2009). Tonnage for selected U.S. ports in 2009. <http://www.ndc.iwr.usace.army.mil//wcsc/portname09.htm>

[30] National Oceanic and Atmospheric Administration. (1975). *The coastline of the United States.* <http://shoreline.noaa.gov/_pdf/ Coastline_of_the_US_1975.pdf> Includes length of general outline of seacoast; does not include freshwater coastlines.

[31] National Oceanic and Atmospheric Administration. (1975). *The coastline of the United States.* <http://shoreline.noaa.gov/_pdf/Coastline_of_the_US_1975.pdf> Includes length of outer coast shoreline, offshore islands, sounds, bays, rivers, and creeks to head of tidewater.

[32] U.S. Census Bureau. (2011). Census 2010.<http://factfinder2.census.gov/faces/nav/jsf/pages/index.xhtml>

[33] U.S. Census Bureau. (2011). Census 2010.<http://factfinder2.census.gov/faces/nav/jsf/pages/index.xhtml> Includes watershed counties that NOAA indicates have either (1) at least 15 percent of their land area within a coastal watershed, or (2) a portion of or an entire county accounts for at least 15 percent of a U.S. Geological Survey coastal cataloging unit.

[34] Florida Department of Environmental Protection GIS analysis coastal county shoreline data (1:40,000) and state offshore territorial boundary (November 2011).

[35] VISIT FLORIDA. (2011). 2010-2011 Annual Report. <http://www. visitflorida.org/am/vfcustom/annualreport/VF_annualreport2010_125.html>

[36] Florida Fish and Wildlife Conservation Commission. (2011). *Economics of fish and wildlife recreation: Seafood industry and boating*. <http:// myfwc.com/media/1496521/EconomicsOfFishAndWildlife-July2011. pdf>

[37] Southwick, R. & Allen, T. (2008). *The 2006 economic benefits of wildlife viewing recreation in Florida*. Report to the Florida Fish and Wildlife Conservation Commission.<http://myfwc.com/media/131044/ WldlfViewing_economics_report.pdf>

[38] Florida Ports Council. (2011). *Florida seaports: A dynamic economic system*. <http://www.epa.gov/gulfcoasttaskforce/pdfs/1110importanceofflorida.pdf> (As cited in Martin Associates. 2009. Statewide economic impacts of maritime cargo handled at Florida's public seaports—2008. Final report to the Florida Ports Council.)

[39] University of West Florida. (2010). *Florida defense industry economic impact analysis*, January 2010—volume *1*. <http://www.pagesinventory.com/visit.php?domain=www.floridadefense.org>

[40] National Oceanic and Atmospheric Administration. (2011). *Fisheries economics of the United States*: 2009. NOAA Technical Memorandum NMFS-F/SPO-109. Accessed September 2011.<http://www.st.nmfs.noaa.gov/st5/publication/econ/2009/FEUS%20 2009%20ALL.pdf>

[41] National Oceanic and Atmospheric Administration. (2011). Fisheries economics of the United States: 2009. NOAA Technical Memorandum NMFS-F/SPO-109. Accessed September 2011. <http://www.st.nmfs.noaa.gov/st5/publication/econ/2009/FEUS%20 2009%20ALL.pdf>

[42] Florida Fish and Wildlife Conservation Commission. (2011). *The economic impact of saltwater fishing in Florida*. <http://www.myfwc.com/conservation/value/saltwater-fishing/>

[43] Florida Fish and Wildlife Conservation Commission. (2011). *The economic impact of saltwater fishing in Florida*. <http://www.myfwc. com/conservation/value/saltwater-fishing/>

[44] National Oceanic and Atmospheric Administration. (1975). *The coastline of the United States*. <http://shoreline.noaa.gov/_pdf/Coastline_of_the_US_1975.pdf> Includes length of general outline of seacoast; does not include freshwater coastlines.

[45] National Oceanic and Atmospheric Administration. (1975). *The coastline of the United States*. <http://shoreline.noaa.gov/_pdf/Coastline_of_the_US_1975.pdf> Includes length of outer coast shoreline, offshore islands, sounds, bays, rivers, and creeks to head of tidewater.

[46] U.S. Census Bureau. (2011). Census 2010. <http://factfinder2.census.gov/faces/nav/jsf/pages/index.xhtml>

[47] U.S. Census Bureau. (2011). Census 2010. <http://factfinder2.census.gov/faces/nav/jsf/pages/index.xhtml> Includes watershed counties that NOAA indicates have either (1) at least 15 percent of their land area within a coastal watershed, or (2) a portion of or an entire county accounts for at least 15 percent of a U.S. Geological Survey coastal cataloging unit.

[48] Coastal Protection and Restoration Authority of Louisiana. (2011). Fiscal year 2012 annual plan. <http://www.lacpra.org/assets/docs/2012%20ANNUAL%20PLAN/ CPRA _Annual_Plan_4-28-11_Web_Format. pdf> p. viii.

[49] U.S. Army Corps of Engineers. (2010). CY 2009 waterborne tonnage by state (in units of 1,000 tons).<http://www.ndc.iwr.usace.army.mil//wcsc /statenm09.htm>

[50] U.S. Army Corps of Engineers. (2009). Tonnage for selected U.S. ports in 2009. <http://www.ndc.iwr.usace.army.mil//wcsc/portname09.htm>

[51] Cieslak, V. (2005). *Ports in Louisiana: New Orleans, South Louisiana, and Baton Rouge.* CRS Report for Congress RS22297. <http://fpc.state.gov/documents/organization/57872.pdf>

[52] Louisiana Department of Natural Resources. (2011). *Louisiana energy facts: Annual 2010.*<http://dnr.louisiana.gov/assets/TAD/newsletters/energy_facts_annual/LEF_2010.pdf> p. E-17.

[53] Louisiana Department of Natural Resources, Technology Assessment Division. (2010). *Selected Louisiana energy statistics.* <http://dnr.louisiana.gov/assets/docs/energy/newsletters/2010/2010-06_topic_1. pdf>

[54] Office of Natural Resources Revenue. (2010). *Total federal offshore reported royalty revenues.* <http://www.onrr.gov/ONRRWebStats/FedOffReportedRoyaltyRevenues.aspx?yeartype=FY&year=2008&datetype=AY>

[55] National Marine Fisheries Service. (2010). *Fisheries of the United States*—2010. <http://www.st.nmfs.noaa.gov/st1/fus/fus10/index.html>

[56] National Oceanic and Atmospheric Administration. (2010). *Shorelines and coastal habitats in the Gulf of Mexico*. <http://gulfseagrant.tamu.edu/oilspill/pdfs/Shorelines_coastal_habitats_FACT_SHEET.pdf>

[57] Louisiana Department of Culture, Recreation and Tourism. n.d. Louisiana tourism by the numbers. <http://www.crt.state.la.us/tourism/research/Documents/2010-11/Louisiana TourismFactsUpdatedfullsheet. pdf>

[58] National Oceanic and Atmospheric Administration. (1975). *The coastline of the United States.* <http://shoreline.noaa.gov/_pdf/Coastline_of_the_US_1975.pdf> Includes length of general outline of seacoast; does not include freshwater coastlines

[59] National Oceanic and Atmospheric Administration. (1975). *The coastline of the United States.* <http://shoreline.noaa.gov/_pdf/Coastline_of_the_US_1975.pdf> Includes length of outer coast shoreline, offshore islands, sounds, bays, rivers, and creeks to head of tidewater.

[60] Google Maps. (2011). Straight line mileage measured from the Mississippi-Louisiana state line at the mouth of the Pearl River to the Mississippi-Alabama state line at Grand Bay National Estuarine Research Reserve.

[61] U.S. Census Bureau. (2011). Census 2010. <http://factfinder2.census.gov/faces/nav/jsf/pages/index.xhtml>

[62] U.S. Census Bureau. (2011). Census 2010. <http://factfinder2.census.gov/faces/nav/jsf/pages/index.xhtml> Includes watershed counties that NOAA indicates have either (1) at least 15 percent of their land area within a coastal watershed, or (2) a portion of or an entire county accounts for at least 15 percent of a U.S. Geological Survey coastal cataloging unit.

[63] Mississippi Department of Environmental Quality. (2010). State of Mississippi water quality assessment 2010 Section 305(b) report. <http://www.deq.state.ms.us/mdeq.nsf/pdf/FS_MS_2010_305_b_report/$File/MS_2010_305_b__Report.pdf?OpenElement>

[64] Mississippi Department of Environmental Quality. (2008). *Citizen's guide to water quality in the Coastal Streams Basin.* <http://www.deq.state.ms.us/mdeq.nsf/pdf/WMB_CoastalCitizenGuide112008/$File/Coastal%20St%20Cit%20Guide.pdf?OpenElement>p. 11.

[65] Mississippi Development Authority, Tourism Division. (2011). 2010 *travel and tourism economic contribution report.* <http://www. visitmississippi.org/uploads/docs/reports_statistics/Economic%20Contribution%20Report%20FY%202010.pdf>

[66] Ingalls Shipbuilding. (2011). Welcome. <http://www. huntingtoningalls.com/is/>

[67] National Oceanic and Atmospheric Administration. (1975). *The coastline of the United States.* <http://shoreline.noaa.gov/_pdf/Coastline_of_the_US_1975.pdf> Includes length of general outline of seacoast; does not include freshwater coastlines.

[68] National Oceanic and Atmospheric Administration. (1975). *The coastline of the United States.* <http://shoreline.noaa.gov/_pdf/Coastline_of_the_US_1975.pdf> Includes length of outer coast shoreline, offshore islands, sounds, bays, rivers, and creeks to head of tidewater.

[69] U.S. Census Bureau. (2011). Census 2010. <http://factfinder2.census.gov/faces/nav/jsf/pages/index.xhtml> Includes watersheds that NOAA indicates have either (1) at least 15 percent of their land area within a coastal watershed, or (2) a portion of or an entire county accounts for at least 15 percent of a U.S. Geological Survey coastal cataloging unit.

[70] U.S. Census Bureau. (2011). Census 2010. <http://factfinder2.census.gov/faces/nav/jsf/pages/index.xhtml> Includes watersheds that NOAA indicates have either (1) at least 15 percent of their land area within a coastal watershed, or (2) a portion of or an entire county accounts for at least 15 percent of a U.S. Geological Survey coastal cataloging unit.

[71] U.S. Department of Transportation. (2009). *Freight facts and figures* 2007. <http://www.ops.fhwa.dot.gov/freight/freight_analysis/nat_freight_stats/docs/07factsfigures/table2_1.htm>

[72] Texas Department of Transportation. n.d. Gulf intracoastal waterway. <http://ftp.dot.state.tx.us/pub/txdotinfo/library/reports/gov/tpp/giww08.pdf> p. 5.

[73] Redwine, A. *The economic value of the Texas Gulf Coast.* <http://gbic.Tamug.edu/gbeppubs/T1/gbnepT1_01-06.pdf> p. 3.

[74] U.S. Army Corps of Engineers. (2010). *Tonnage for Selected U.S. Ports in 2009.* <http: //www.ndc.iwr.usace.army.mil//wcsc/portname09.htm>

[75] National Petrochemical and Refiners Association. (2004). *NPRA United States refining and storage capacity.* <http://www.npra.org/docs/publications/statistics/RC2004.pdf> p. 2.

[76] Combs, S. (2007). Window on state government. <http://www.window.state.tx.us/specialrpt/energy/nonrenewable/crude.php>
[77] National Oceanic and Atmospheric Administration. (2009). *Regional summary Gulf of Mexico region management context.* <http://www.st.nmfs.noaa.gov/st5/publication/econ/2009/gulf_summary_econ.pdf>
[78] Southwick Associates. (2007). *Sportfishing in America—an economic engine and conservation powerhouse*. Produced for the American Sportfishing Association with funding from the Multistate Conservation Grant Program.
[79] Center for Texas Beaches and Shores. (2005). *The dynamic Texas coast.* <http://www.tamug.edu/CTBS/about_us/historymission/doc/Texas%20Coast%20Powerpoint.pdf>
[80] Center for Texas Beaches and Shores. (2005). *The dynamic Texas coast*. <http://www.tamug.edu/CTBS/about_us/historymission/doc/Texas%20Coast%20 Powerpoint.pdf>
[81] Texas Department of Economic Development—(2002) data.
[82] U.S. Army Corps of Engineers. (2011). *Dredging program.* <http://www.ndc.iwr.usace.army.mil/dredge/dredge.htm>
[83] U.S. Geological Survey. (2003). *Without restoration, coastal land loss to continue.* Accessed September 23, 2011. <http://www.nwrc.usgs. gov/ releases/pr03_004.htm>
[84] Couvillion, B. R., Barras, J. A., Steyer, G. D., Sleavin, W., Fischer, M., Beck, H., Trahan, N., Griffin, B. & Heckman, D. (2011). Land area change in coastal Louisiana from 1932 to 2010. U.S. Geological Survey Scientific Investigations Map 3164. <http://pubs.usgs.gov/sim/3164/>
[85] Couvillion, B. R., Barras, J. A., Steyer, G. D., Sleavin, W., Fischer, M., Beck, H., Trahan, N., Griffin, B. & Heckman. D. (2011). Land area change in coastal Louisiana from 1932 to 2010. U.S. Geological Survey Scientific Investigations Map 3164. <http://pubs.usgs.gov/sim/3164/>
[86] Couvillion, B. R., Barras, J. A., Steyer, G. D., Sleavin, W., Fischer, M., Beck, H., Trahan, N., Griffin, B. & Heckman. D. (2011). Land area change in coastal Louisiana from 1932 to 2010. U.S. Geological Survey Scientific Investigations Map 3164. <http://pubs.usgs.gov/sim/3164/>
[87] U.S. Environmental Protection Agency. (2008). *Nitrogen and phosphorus loads in large rivers. In: Report on the environment*. <http:// cfpub.epa.gov/eroe/index.cfm?fuseaction=detail.viewInd&lv=list.listBySubTopic&r=216594&subtop=200&ch=47> Original data provided to ERG (an EPA contractor) by Nancy Baker, USGS. September 12, 2007.
[88] Doney, S. C., Mahowald, N., Lima, I., Feely, R. A., Mackenzie, F. T., Lamarque, J. -F. & Rasch, P. J. (2007). *Impact of anthropogenic atmospheric nitrogen and sulfur deposition on ocean acidification and the inorganic carbon system*. Proc. Natl. Acad. Sci. U S A 104(37): 14580-14585. <http://www.pnas.org/content/104/37/14580.full.pdf+html>
[89] Mississippi River/Gulf of Mexico Watershed Nutrient Task Force. (2008). *Gulf hypoxia action plan 2008 for reducing, mitigating, and controlling hypoxia in the northern Gulf of Mexico and improving water quality in the Mississippi River Basin*. Accessed September 23, 2011. <http://www.epa.gov/owow/msbasin/pdf/ghap2008_update 082608.pdf> p. 9.
[90] Rabalais, N. N. & Turner, R. E. (2011). Louisiana Universities Marine Consortium. Shelfwide cruise <http://www.gulfhypoxia.net/research/ Shelfwide%20Cruises/2011/>

[91] Mississippi River/Gulf of Mexico Watershed Nutrient Task Force. (2008). Gulf hypoxia action plan 2008 for reducing, mitigating, and controlling hypoxia in the northern Gulf of Mexico and improving water quality in the Mississippi River Basin. Accessed September 23, 2011. <http://www.epa.gov/owow/msbasin/pdf/ghap2008_up date082608.pdf> p.32.

[92] National Resources Conservation Service. (2011). *An assessment of the effects of conservation practices on cultivated cropland in the Great Lakes region.* <http://www. nrcs.usda.gov/Internet/FSE_ DOCUMENTS/ stelprdb1045481.pdf>

[93] U.S. Environmental Protection Agency. (2010). *Designation of North American Emission Control Area to reduce emissions from ships: Regulatory announcement.* EPA-420-F-10-015.<http://www.epa.gov/otaq/regs/nonroad/marine/ci/420f10015.htm>

[94] U.S. Environmental Protection Agency. (2010). *U.S.-Mexico demonstration of fuel switching on ocean going vessels in the Gulf of Mexico.* EPA-160-R-10-001. <http:// www.epa.gov/international/fuel switch.html>

[95] U.S. Fish and Wildlife Service and U.S. Census Bureau. (2006). 2006 national survey of fishing, hunting, and wildlife-associated recreation. <http://www.census.gov/prod/ 2008pubs/fhw06-nat.pdf> Table 52.

[96] Southwick, R., and Allen, T. (2008). The 2006 economic benefits of wildlife viewing recreation in Florida. <http://floridabirdingtrail.com /images/pages/wv_economics_ report.pdf> Table 3.

[97] National Marine Fisheries Service. (2010). Annual commercial landings statistics. Years queried: 2007-2009. Accessed October 4, 5, and 6, 2010, and November 19, 2010. (As cited in National Oceanic and Atmospheric Administration. 2011. The Gulf of Mexico at a glance: A second glance. <http://stateofthecoast.noaa.gov/NOAAs_ Gulf_of_ Mexico_at_a_Glance_report.pdf>)

[98] National Marine Fisheries Service. Fisheries of the United States—(2009). <http:// www.st.nmfs.noaa.gov/st1/fus/fus09/index.html>

[99] National Oceanic and Atmospheric Administration. (2011). *The Gulf of Mexico at a glance: A second glance.* <http://stateofthecoast. noaa.gov/NOAAs_Gulf_of_Mexico_ at_a_Glance_report.pdf>

[100] U.S. Fish and Wildlife Service. (2011). Endangered Species Act Program. <http://www. fws.gov/endangered/>

[101] SouthEast Data, Assessment, and Review. (2009). *Stock assessment of red snapper in the Gulf of Mexico—SEDAR update assessment.* <http://www.sefsc.noaa.gov/sedar/ download/Red%20Snapper%20Update%202009%205.0.pdf?id=DOCUMENT>

[102] Magnuson-Stevens Fisheries Conservation and Management Act. (1996). Section 407. <http://www.nmfs.noaa.gov/sfa/magact/mag4. html>

[103] National Marine Fisheries Service, U.S. Fish and Wildlife Service, and SEMARNAT. (2001). Bi-national recovery plan for the Kemp's Ridley sea turtle (*Lepidochelys kempii*). Second revision. <http://www.nmfs. noaa.gov/pr/pdfs/recovery /kempsridley_ revision2.pdf>

[104] U.S. Fish and Wildlife Service. (2011). *Whooping cranes return celebrated at White Lake Wetlands Conservation Area.* <http://www.fws.gov/southeast/news/2011/11-017. html>

[105] U.S. Geological Survey. (2010). *Nonindigenous Aquatic Species* (NAS) database query. <http://nas.er.usgs.gov/queries/stco.aspx>

[106] U.S. Department of Agriculture. (2011). *Species profiles: Nutria.* <http://www. invasivespeciesinfo.gov/aquatics/nutria.shtml>

[107] Schofield, P. J. (2010). Update on geographic spread of invasive lionfishes (*Pterois volitans* [Linnaeus, 1758] and *P. miles* [Bennett, 1828]) in the Western North Atlantic Ocean, Caribbean Sea and Gulf of Mexico. Aquat. Invasions 5(1): S117-S122. <http:// www.aquatic invasions.net/2010/Supplement/AI_2010_5_S1_Schofield>

[108] U.S. Department of Agriculture. (2011). *National Invasive Species Information Center (NISIC): Gateway to invasive species information; covering federal, state, local, and international sources.* <http://www. invasivespeciesinfo.gov/index.shtml>

[109] National Marine Sanctuaries. (2008). *Flower Garden Banks: 2008 condition report.* <http://sanctuaries.noaa.gov/science/condition/fgbnms/ pressures.html>

[110] Houston Advanced Research Center. (2010). *Blue shrimp, Pacific white shrimp, Asian tiger shrimp.* <http://www.galvbayinvasives.org/Guide/Species/LitopenaeusMacro brachiumPenaeus>

[111] McGuire, M. & Stevely, J. (2009). *Invasive species of Florida's coastal waters: The Asian green mussel* (*Perna viridis*). <http://edis.ifas.ufl.edu/sg094>

[112] National Park Service. n.d. Tilapia zilli. <http://www.nature.nps.gov/water/marine invasives/assets/PDFs/Tilapia_zilli.pdf>

[113] NOAA Research. (2011). *Aquatic invasive species.* <http://www.oar.noaa.gov/oceans/t _invasivespecies.html>

[114] Pimentel, D., Zuniga, R. & Morrison. D. (2005). Update on the environmental and economic costs associated with alien-invasive species in the United States. *Ecol. Econ. 52*: 273-288.

[115] Blake, E. S., & Gibney, E. J. The deadliest, costliest, and most intense United States tropical cyclones from 1851 to 2010 (and other frequently requested hurricane facts). NOAA Technical Memorandum NWS NHC-6. <http://www.nhc.noaa.gov/pdf/nws-nhc -6.pdf> p. 9.

[116] Congressional Budget Office. (2007). *The federal government's spending and tax actions in response to the Gulf Coast Hurricanes.* <http://www.cbo.gov/ftpdocs/85xx/ doc8514/08-07-Hurricanes_Letter. pdf>

[117] Multihazard Mitigation Council. (2005). *Natural hazard mitigation saves: An independent study to assess the future savings from mitigation activities.* Volume *1*: Findings, conclusions, and recommendations. National Institute of Building Sciences. <http://www.nibs.org/client/ assets/files/mmc/Part1_final.pdf>

[118] Louisiana Department of Wildlife and Fisheries. (2011). Oil spill response. <http:// www.wlf.louisiana.gov/oilspill>

[119] Levin, P. S., Fogarty, M. J., Matlock, G. C. & Ernst, M. (2008). *Integrated ecosystem assessments.* NOAA Technical Memorandum NMFS-NWFSC-92. <http://www.st. nmfs.noaa.gov/iea/documents/IEA_ TM92Final.pdf>

[120] Levin, P. S., Fogarty, M. J., Murawski, S. A. & Fluharty, D. (2009). *Integrated ecosystem assessments: Developing the scientific basis for ecosystem-based management of the ocean. PLoS Biol.* 7(1):e1000014.doi:10.1371/journal.Pbio. 1000014. <http://www.plosbiology.org/article/info:doi/10.1371/journal.pbio.1000014 >

[121] Alabama Department of Conservation and Natural Resources, State Lands Division. (1979). Alabama Coastal Area Management Program.

[122] Alabama Department of Environmental Management.

[123] U.S. Census Bureau. n.d. State and county quickfacts. <http:// quickfacts.census.gov>

[124] The Nature Conservancy. (2002). *States of the union: Ranking America's biodiversity.* <http://www.natureserve.org/library/stateof unions.pdf>

[125] Union of Concerned Scientists. (2009). Alabama. <http:// www. ucsusa.org/gulf/ gcstateala_bio.html>

[126] Smith, W. E. (1988). *Geomorphology of the Mobile delta. Geological Survey of Alabama Bulletin* #132. <http://digital.archives. alabama.gov/cdm4/item_viewer.php? CISOROOT=/ADAH&CISOPTR=509&CISOBOX=1&REC=15> pp. 1, 6.

[127] Rainer, D. (2008). *Alliance's goal: Healthy Gulf of Mexico.* <http://www. outdooralabama.com/oaonline/gulfalliance09.cfm>

[128] Mississippi-Alabama Sea Grant Consortium. n.d. MASGC focus on habitat restoration. <http://www.masgc.org/page.asp?id=219>

[129] Alabama Department of Tourism. (2010). *Travel Economic Impact Report* 2009. <http://www.alabama.travel/media/media_room/Report/ 2009TourismReport.pdf>

[130] U.S. Army Corps of Engineers. (2009). Tonnage for selected U.S. ports in 2009. <http://www.ndc.iwr.usace.army.mil//wcsc/portname09.htm>

[131] Alabama State Port Authority. (2010). *Alabama State Port Authority dedicates new turning basin.* <http://www.asdd.com/pdf/ PortofMobile _TurningBasin_Dedicated_ 09092010.pdf>

[132] Alabama State Port Authority. (2011). *Port facts.* <http://www.asdd. com>

[133] Mobile Chamber of Commerce. (2010). *An economic overview of the Mobile Bay region.* <http://www.mobilechamber.com/ regional overview.pdf> p. 7.

[134] Mobile Chamber of Commerce. (2010). *An economic overview of the Mobile Bay region.* <http://www.mobilechamber. com/regional overview.pdf> p. 47.

[135] Alabama Department of Conservation and Natural Resources. Annual beach profiles and analyses, submitted by Geological Survey of Alabama.

[136] Alabama Department of Conservation and Natural Resource, Parks Division; Alabama Department of Conservation and Natural Resources, State Lands Division.

[137] Alabama Department of Insurance. (2011). *Tornadoes' insured losses likely to exceed Ivan.* <http://www.aldoi.gov/current newsitem. aspx?ID=776>

[138] National Oceanic and Atmospheric Administration. (1975). *The coastline of the United States.* <http://shoreline.noaa.gov/_pdf/Coastline_of_the_US_1975.pdf> Includes length of outer coast shoreline, offshore islands, sounds, bays, rivers, and creeks to head of tidewater.

[139] Florida Department of Environmental Protection. n.d. Florida shoreline length information according to beaches and shores technical and design memorandum 89-1, 5th Edition, December 1993, entitled beach conditions in Florida: A statewide inventory and identification of the beach erosion problem areas in Florida by Ralph R. Clark. <http://www.dep.state.fl.us/beaches/publications/pdf/fl_beach.pdf>

[140] Florida Department of Environmental Protection GIS analysis coastal county shoreline data (1:40,000) and state offshore territorial boundary (November 2011).

[141] U.S. Census Bureau. (2011). Census 2010. <http://factfinder2.census. gov/faces/nav/ jsf/pages/index.xhtml>

[142] Dr. Beach. (2010). *The best beaches in America.* <http://www.drbeach.org/top10 beaches.htm>

[143] VISIT FLORIDA. (2011). 2010-2011 annual report. <http://www. visitflorida.org/am/vfcustom/annualreport/VF_annualreport2010_125.html>

[144] Florida Fish and Wildlife Conservation Commission. (2011). *Economics of fish and wildlife recreation: Seafood industry and boating*. <http:// myfwc.com/media/1496521/EconomicsOfFishAndWildlife-July2011. pdf>

[145] Southwick, R. & Allen, T. (2008). *The 2006 economic benefits of wildlife viewing recreation in Florida.* Report to the Florida Fish and Wildlife Conservation Commission. <http://myfwc.com/media/131044/WldlfViewing_economics_report.pdf>

[146] Harper, R. Pooley, P. K. & Scheibe, M. (2011). *Florida defense industry economic impact analysis.* Final report. Volume *1*. <http://www.floridadefense.org/documents/HAAS%20Study%202011/FLdefense_ Volume_1.pdf>

[147] Harper, R., Pooley, P. K. & Scheibe, M. (2011). *Florida defense industry economic impact analysis.* Final report. Volume *1*. <http://www.floridadefense.org/documents/HAAS%20Study%202011/FLdefense_Volume_1.pdf>

[148] Florida Ports Council. (2011). *Florida seaports: A dynamic economic system.* <http://www.epa.gov/gulfcoasttaskforce/pdfs/1110importanceof florida.pdf> (As cited in Martin Associates. 2009. Statewide economic impacts of maritime cargo handled at Florida's public seaports—2008.)

[149] Florida Fish and Wildlife Conservation Commission. (2011). *The economic impact of saltwater fishing in Florida.* <http://www.myfwc. com/conservation/value/saltwater-fishing/>

[150] Florida Fish and Wildlife Conservation Commission. (2011). *The economic impact of saltwater fishing in Florida.* <http://www.myfwc. com/conservation/value/saltwater-fishing/>

[151] Florida Fish and Wildlife Conservation Commission. (2011). *The economic impact of saltwater fishing in Florida.* <http://www.myfwc. com/conservation/value/saltwater-fishing/>

[152] National Oceanic and Atmospheric Administration. (2011). Fisheries economics of the United States: 2009. NOAA Technical Memorandum NMFS-F/SPO-109. Accessed September 2011. <http://www.st.nmfs.noaa.gov/st5/publication/econ/2009/FEUS%20 2009%20ALL.pdf>

[153] National Oceanic and Atmospheric Administration. (2011). Fisheries economics of the United States: 2009. NOAA Technical Memorandum NMFS-F/SPO-109. Accessed September 2011. <http://www.st.nmfs.noaa.gov/st5/publication/econ/2009/FEUS%20 2009%20ALL.pdf>

[154] Couch, C., Hopkins, E. Hardy, P. (2010). *Influences of environmental settings on aquatic ecosystems in the Apalachicola-Chattahoochee-Flint River Basin.* 2010. USGS Water-Resources Investigations Report 95-4287. <http://pubs.usgs.gov/wri/1995/4278/report.pdf>

[155] Florida Institute of Oceanography. (2010). Members. <http://www.fio.usf.edu/WhoWe Are/Members.aspx>

[156] Florida Department of Environmental Protection. (2010). Integrated water quality assessment for Florida: 2010 305(b) report and 303(d) list update. <http://www.dep. state.fl.us/water/docs/2010_Integrated_Report. pdf> p. 129.

[157] Federal Emergency Management Agency. (2011). Florida disaster history. <http:// www.fema.gov/news/disasters_state.fema?id=12>

[158] Louisiana State University AgCenter. n.d. Portrait of an estuary: Functions and values of the Barataria-Terrebonne estuary system. <http://www.lsu.edu/seagrantfish/pdfs/portrait_estuary.pdf> p. 3.

[159] Couvillion, B. R., Barras, J. A., Steyer, G. D., Sleavin, W., Fischer, M., Beck, H., Trahan, N. Griffin, B., & Heckman, D. (2011). Land area change in coastal Louisiana from 1932 to 2010: U.S. Geological Survey Scientific Investigations Map 3164. <http://pubs.usgs.gov/sim/3164>

[160] National Oceanic and Atmospheric Administration. (2009). Fishing communities of the United States: 2006. NOAA Technical Memorandum NMFS-F/SPO-98. <http://www.st.nmfs.noaa. gov/st5/ publication/fisheries_communities_2006.html>

[161] Louisiana Department of Natural Resources, Technology Assessment Division. (2010). *Louisiana energy facts*. <http://dnr.louisiana.gov/assets/TAD/newsletters/energy_facts _annual/LEF_2010.pdf>

[162] Coastal Protection and Restoration Authority. (2011). *Fiscal year 2012 annual plan: Integrated ecosystem restoration and hurricane protection in Coastal Louisiana.* Coastal Protection and Restoration Authority of Louisiana. <http://www.lacpra.org/assets/docs/2012%20ANNUAL%20PLAN/CPRA_Annual_Plan_4-28-11_Web_ Format.pdf>

[163] Couvillion, B. R., Barras, J. A., Steyer, G. D., Sleavin, W. Fischer, M., Beck, H., Trahan, N. Griffin, B. & Heckman, D. (2011). Land area change in coastal Louisiana from 1932 to 2010: U.S. Geological Survey Scientific Investigations Map 3164. <http://pubs.usgs.gov/sim/3164>

[164] Coastal Protection and Restoration Authority of Louisiana. (2011). Fiscal year 2012 annual plan. <http://www.lacpra.org/assets/docs/2012%20ANNUAL%20PLAN/CPRA _Annual_Plan_4-28-11_Web_ Format.pdf> p. viii.

[165] Couvillion, B. R., Barras, J. A., Steyer, G. D., Sleavin, W., Fischer, M., Beck, H., Trahan, N., Griffin, B. & Heckman, D. (2011). Land area change in coastal Louisiana from 1932 to 2010: U.S. Geological Survey Scientific Investigations Map 3164. <http://pubs.usgs.gov/sim/3164>

[166] Congressional Budget Office. (2007). The federal government's spending and tax actions in response to the Gulf Coast Hurricanes. <http://www.cbo.gov/ftpdocs/85xx/doc8514/08-07Hurricanes_Letter. pdf>

[167] Scott, L. (2011). *The energy sector: Still a giant economic engine for the Louisiana economy.* Mid-Continent Oil and Gas Association. <http://the hayride.com/wp-content/uploads/2011/10/Executive-Summary-1.pdf>

[168] Louisiana Mid-Continent Oil and Gas Association. n.d. Louisiana oil and gas pipelines. <http://www.lmoga.com/industry-sectors/pipelines/>

[169] Louisiana Department of Natural Resources. n.d. America's energy corridor: Louisiana serving the nation's energy needs. <http://dnr. louisiana.gov/assets/docs/ energy/policy papers/AW_AmericasEnergyCorridor_Revised.pdf> p. 18.

[170] Scott, L. C. (2008). The economic impacts of Port Fourchon on the national and Houma MSA economies. <http://www.portfourchon.com/site10001/1001757/docs/port_ four chon_economic_impact_study.pdf>

[171] Scott, L. C. (2008). *The economic impacts of Port Fourchon on the national and Houma MSA economies.* <http://www.portfourchon.com/site100-01/1001757/docs/port _fourchon_ economic_impact_study. pdf>

[172] Scott, L. C. (2008). *The economic impacts of Port Fourchon on the national and Houma MSA economies.* <http://www.portfourchon.com/site100-01/1001757/docs/port_fourchon_economic _impact_study.pdf>
[173] U.S. Energy Information Administration. n.d. Louisiana 2009: *Distribution of wells by production rate bracket.* <http://www.eia. gov/pub/ oil_gas/petrosystem/la_table.html>
[174] U.S. Department of Energy. (2011). *Strategic petroleum reserve storage sites.* <http://www.fe.doe.gov/programs/reserves/spr/spr-sites.html>
[175] U.S. Army Corps of Engineers. (2010). Tonnage for selected U.S. ports in 2009. <http://www.ndc.iwr.usace.army.mil/wcsc/portton09.htm>
[176] U.S. Army Corps of Engineers. (2010). CY 2009 waterborne tonnage by state (in units of 1,000 tons). <http://www.ndc.iwr.usace.army. mil//wcsc/statenm09.htm>
[177] U.S. Army Corps of Engineers. (2010). CY 2009 waterborne tonnage by state (in units of 1,000 tons). <http://www.ndc.iwr.usace.army. mil// wcsc/statenm09.htm>
[178] State of Louisiana. Office of the Governor. (2011). *Louisiana's coast: Ecosystem restoration and flood protection.* <http://coastal.louisiana.gov/index.cfm?md=pagebuilder&tmp=home&pid=115>
[179] Cieslak, V. (2005). *Ports in Louisiana: New Orleans, South Louisiana, and Baton Rouge.* CRS Report for Congress RS22297. <http://fpc.state.gov/documents/organization/57872.pdf>
[180] National Marine Fisheries Service. (2010). Fisheries of the United States—2010. <http://www.st.nmfs.noaa.gov/st1/fus/fus10/index.html>
[181] National Marine Fisheries Service. (2010). Annual commercial landing statistics. <http://www.st.nmfs.noaa.gov/st1/commercial/landings/annual _landings.html>
[182] National Marine Fisheries Service. n.d. U.S. commercial landings. <http://www.st.nmfs.noaa.gov/st1/fus/fus10/02_commercial2010.pdf>p. 7.
[183] Southwick Associates, Inc. (2008). *The economic benefits of fisheries, wildlife and boating resources in the state of Louisiana*—2006. <http://www.wlf.louisiana.gov/sites/default/files/pdf/publication/32728-economic-benefits-fisheries-wildlife-and-boating-resources-state-louisiana-2006/southwick_2006_final_final_report_5-27-08_ 0. pdf>
[184] Louisiana Department of Wildlife and Fisheries. (2011). *Waterfowl population estimates in Louisiana's coastal zone below U.S. Highway 90 and on Catahoula Lake.* <http://www.wlf.louisiana.gov/sites/default/files/pdf/waterfowl_survey/33575-January%202011%20Survey/ waterjan2011.pdf>
[185] U.S. Fish and Wildlife Service. (2011). *Listings and occurrences for Louisiana.* <http://ecos.fws.gov/tess_public/pub/stateListingAndOccurrenceIndividual.jsp?state=LA>
[186] National Oceanic and Atmospheric Administration. (2011). *NOAA's state of the coast.* <http://stateofthecoast.noaa.gov/hypoxia/welcome. html>
[187] Coastal Protection and Restoration Authority. (2007). *Integrated ecosystem restoration and hurricane protection: Louisiana's comprehensive master plan for a sustainable coast.* <http://coastal.louisiana.gov/index.cfm?md=pagebuilder&tmp=home&nid=24&pnid=0&pid=28&fmid=0&catid=0&elid=0>
[188] U.S. Census Bureau. (2010). *Percent of the native population born in their state of residence (including Puerto Rico).* <http://factfinder2.census.gov/faces/tableservices/jsf/pages/productview.xhtml?pid=ACS_10_1YR_GCT0601.US01PR&prodType=table>

[189] Barras, J. A. (2009). *Land area change and overview of major hurricane impacts in coastal Louisiana,* 2004-08. U.S. Geological Survey Scientific Investigations Map 3080. <http://pubs.usgs.gov/sim/3080/>

[190] Coastal Protection and Restoration Authority. (2009). *Fiscal year 2010 annual plan: integrated ecosystem restoration and hurricane protection in coastal Louisiana.* Coastal Protection and Restoration Authority of Louisiana. <http://www.lacpra.org/assets/docs/FY2010%20Annual%20 Plannew.pdf>

[191] Couvillion, B. R., Barras, J. A., Steyer, G. D., Sleavin, W., Fischer, M., Beck, H., Trahan, N., Griffin, B. & Heckman, D. (2011). Land area change in coastal Louisiana from 1932 to 2010: U.S. Geological Survey Scientific Investigations Map 3164. <http://pubs.usgs.gov/sim/3164>

[192] Louisiana Department of Natural Resources, Technology Assessment Division. (2010). *Louisiana energy facts.* <http://dnr.louisiana.gov/assets/TAD/newsletters/energy_facts_annual/LEF_2010.pdf>

[193] U.S. Energy Information Administration. (2011). *Number and capacity of petroleum refineries.* <http://www.eia.gov/dnav/pet/ pet_pnp_cap1_ dcu_ SLA_a.htm>

[194] Louisiana Oil and Gas Association. (2008). LOGA frequently asked questions. <http://www.loga.la/loga-faq.html>

[195] Louisiana Chemical Association. (2011). *Louisiana Chemical Industry Alliance.* <http://www.lca.org/AM/Template.cfm?Section=LCIA>

[196] U.S. Army Corps of Engineers. (2010). CY 2009 waterborne tonnage by state (in units of 1,000 tons). <http://www.ndc.iwr.usace. army.mil// wcsc/statenm09.htm>

[197] U.S. Army Corps of Engineers. (2009). *Waterborne commerce of the United States.* Part 5—national summaries. <http://www.ndc.iwr.usace.army.mil//wcsc/pdf/wcusnatl09.pdf>

[198] U.S. Army Corps of Engineers. (2009). *Waterborne commerce of the United States.* Part 5—national summaries. <http://www.ndc.iwr.usace.army.mil//wcsc/pdf/wcusnatl09.pdf>

[199] State of Louisiana. Office of the Governor. (2011). *Louisiana's coast: ecosystem restoration and flood protection.* <http://coastal.louisiana.gov/index.cfm?md=pagebuilder&tmp=home&pid=112>

[200] National Marine Fisheries Service. (2010). *Fisheries of the United States, Washington, DC.* <http://www.st.nmfs.noaa. gov/st1/fus/fus10/i ndex. html>

[201] Southwick Associates, Inc. (2008). *The economic benefits of fisheries, wildlife and boating resources in the state of Louisiana—2006.* <http://www.wlf.louisiana.gov/sites/default/files/pdf/publication/32728-economic-benefits-fisheries-wildlife-and-boating-resources-state-louisiana-2006/southwick_2006_final_final_report_5-27-08_0.pdf>

[202] Coastal Wetlands, Planning, Protection and Restoration Act. *Louisiana Coastal Wetland Functions and Values.* <http://lacoast.gov/ reports/rtc/1 997/4.htm>

[203] Louisiana Department of Wildlife and Fisheries. (2011). *Waterfowl population estimates in Louisiana's coastal zone below U.S. Highway 90 and on Catahoula Lake.* <http://www.wlf.louisiana.gov/sites/default/files/pdf/waterfowl_survey/33575-January%202011%20Survey/ waterjan2011.pdf>

[204] U.S. Fish and Wildlife Service. (2009). *Southwest Louisiana NWR Complex: General information.* <http://www.fws.gov/ swlarefuge complex/geninfo.html>

[205] U.S. Fish and Wildlife Service. (2011). *Southeast Louisiana National Wildlife Refuges: About the Southeast Louisiana Refuges.* <http://www.fws.gov/southeastlouisiana/about_us.html>

[206] U.S. Census Bureau. (2010). 2010 census data. <http://2010.census.gov/2010census/data>

[207] Louisiana Department of Culture, Recreation and Tourism. n.d. Louisiana tourism by the numbers <http://www.crt.state.la.us/tourism/research/Documents/2010-11/LouisianaTourismFactsUpdatedfullsheet.pdf>

[208] Louisiana Department of Culture, Recreation and Tourism. n.d. Louisiana tourism by the numbers. <http://www.crt.state.la.us/tourism/research/Documents/2010-11/LouisianaTourismFactsUpdated fullsheet.pdf>

[209] Google Maps. (2011). Straight line mileage measured from the Mississippi-Louisiana state line at the mouth of the Pearl River to the Mississippi-Alabama state line at Grand Bay National Estuarine Research Reserve.

[210] U.S. Census Bureau. (2011). Census 2010. <http://factfinder2.census.gov/faces/nav/ jsf/pages/index.xhtml>

[211] Christmas. J. Y. (1973). *Cooperative Gulf of Mexico estuarine inventory and study.* Mississippi. Gulf Coast Research Laboratory.

[212] Mississippi Department of Environmental Quality. (2010). State of Mississippi water quality assessment 2010 Section 305(b) report. <http://deq.state.ms.us/MDEQ.nsf/page/FS_SurfaceWaterQualityAssessments>

[213] Mississippi Department of Environmental Quality. (2008). *Citizen's guide to water quality in the Coastal Streams Basin.* <http://www.deq.state.ms.us/mdeq.nsf/page/WMB_Coastal_Streams_Basin>

[214] Schmidt, K. (2001). Coastal change in Mississippi: A review of 1850-1999 data. *Mississippi Department of Environmental Quality.* <http:// geology.deq.state.ms.us/coastal/NOAA_DATA/Publications/Presentations/Coastwide/CoastwideHistoricalChange.pdf>

[215] Mississippi Department of Marine Resources. (2011). *Artificial reef program.* <http://www.dmr.ms.gov/Fisheries/Reefs/artificial-reefs.htm>

[216] Mississippi Development Authority, Tourism Division. (2005). *Fiscal year 2004 economic impact for tourism in Mississippi.* <http://www.visitmississippi.org/resources/tom_total_fy04_tourism_report.pdf>

[217] Mississippi State Port Authority at Gulfport. n.d. Rebuilding and restoring. <http://www.portofthefuture.com/>

[218] U.S. Air Force. (2009). *81 Training Wing.* <http://www.keesler.af.mil/units/81trainingwing/index.asp>

[219] Ingalls Shipbuilding. (2011). *Welcome.* <http://www.huntingtoningalls. com/is/>

[220] U.S. Census Bureau. (2011). Census 2010. <http://factfinder2.census.gov/faces/nav/jsf/pages/index.xhtml>

[221] Texas Parks and Wildlife. (2010). *Land and water resources conservation and recreation plan.* <http://www.tpwd.state.tx.us/publications/pwdpubs/media/pwd_pl_e0100_867_land_water_plan_01_ 2010.pdf>

[222] Texas Parks and Wildlife. (2010). *Land and water resources conservation and recreation plan.* <http://www.tpwd.state.tx.us/publications/pwdpubs/media/pwd_pl_e0100_867_land_water_plan_01_2010.pdf>

[223] National Climatic Data Center. (2011). *Billion dollar U.S. weather disasters*, 1980-2010. <http://www.ncdc.noaa.gov/ img/reports /billion/ billionz-2010.pdf>

[224] Texas General Land Office. *Coastal dunes: Dune protection and improvement manual for the Texas Gulf Coast.* Fifth edition. <http:// www.glo.texas.gov/what-we-do/caring-for-the-coast/_publications/ DuneManual.pdf>

[225] Texas Coastal Management Program. (2010). *Texas coastal and estuarine land conservation plan.* <http://coastalmanagement.noaa.gov/mystate/docs/celclplantx.pdf> p. 9.

[226] McKenna, K. (2009). *Texas coastwide erosion response plan.* <http://www.glo.texas.gov/what-we-do/caring-for-the-coast/_documents/coastal-erosion/response-plans/coastwide-erosion-response-plan.pdf>p. 4.

[227] Texas Parks and Wildlife Department. <http://www.tpwd.state.tx.us/landwater/water/conservation/txgems/lagmadr/>

[228] Texas Department of Transportation. n.d. Gulf intracoastal waterway. <http://ftp.dot.state.tx.us/pub/txdot-info/library/reports/gov/tpp/giww08. pdf> p. 5.

[229] Texas Department of Transportation. n.d. Gulf intracoastal waterway. <http://ftp.dot.state.tx.us/pub/txdot-info/library/reports/gov/tpp/giww08. pdf> p. 5.

[230] Redwine, A. *The economic value of the Texas Gulf Coast.* <http://gbic.tamug.edu/gbeppubs/T1/gbnepT1_01-06.pdf> p. 3.

[231] U.S. Army Corps of Engineers. (2010). *Tonnage for Selected U.S. Ports in 2009.* <http: //www.ndc.iwr.usace.army.mil//wcsc/portname09.htm>

[232] U.S. Army Corps of Engineers. (2010). *U.S. waterway data: Principal ports of the United States.* Accessed October 2010. <http://www.ndc.iwr.usace.army.mil//data/datappor.htm>

[233] U.S. Army Corps of Engineers. (2010). *Tonnage for Selected U.S. Ports in 2009.* <http: //www.ndc.iwr.usace.army.mil//wcsc/portname09.htm>

[234] U.S. Census Bureau. (2011). Census 2010. <http://factfinder2.census.gov/faces/nav/jsf/pages/index.xhtml>

[235] Port of Houston Authority. (2009). *Overview.* <http://www.portofhouston.com/geninfo/overview1.html>

[236] Port of Houston Authority. (2009). *Overview.* <http://www.portof houston.com/geninfo /overview1.html>

[237] Port of Houston Authority. (2009). *Overview.* <http://www.portofhouston.com/geninfo /overview1.html>

[238] National Petrochemical and Refiners Association. (2004). *NPRA United States refining and storage capacity.* <http://www.npra.org/docs/publications/statistics/RC2004.pdf> p. 2.

[239] Combs, S. (2007). *Window on state government.* <http://www.window.state.tx.us/specialrpt/energy/nonrenewable/crude.php>

[240] National Oceanic and Atmospheric Administration. (2009). *Regional summary Gulf of Mexico region management context.* <http://www.st.nmfs.noaa.gov/st5/publication/econ/2009/gulf_summary_econ.pdf>

[241] National Oceanic and Atmospheric Administration. (2011). *Fisheries economics of the United States: 2009.* NOAA Technical Memorandum NMFS-F/SPO-109. Accessed September 2011. <http://www.st.nmfs.noaa.gov/st5/publication/econ/2009/FEUS%20 2009%20ALL.pdf>

[242] National Oceanic and Atmospheric Administration. (2011). *Texas Commercial Fisheries 2009.* <http://www.st.nmfs.noaa.gov/st5/publication/econ/2009/gulf_TX tables_econ.pdf>

[243] Southwick Associates. (2007). *Sportfishing in America—an economic engine and conservation powerhouse.* Produced for the American Sportfishing Association with funding from the Multistate Conservation Grant Program.

[244] Center for Texas Beaches and Shores. (2005). *The dynamic Texas coast.* <http:// www.tamug.edu/CTBS/about_us/history-mission/doc/Texas%20Coast%20 Powerpoint. pdf>

[245] Center for Texas Beaches and Shores. (2005). *The dynamic Texas coast.* <http://www. tamug.edu/CTBS/about_us/history-mission/doc/Texas%20Coast%20Powerpoint.pdf>

[246] Patterson, J. (2011). *A report to the 82nd Legislature Coastal Erosion and Planning Response Act.* <http://www.glo.texas.gov/what-we-do/caring-for-the-coast/_publica tions/cepra-report-2011.pdf> p. 7.

[247] Patterson, J. (2011). *A report to the 82nd Legislature Coastal Erosion and Planning Response Act.* <http://www.glo.texas.gov/what-we-do/caring-for-the-coast/_publica tions/cepra-report-2011.pdf> p. 7.

[248] Dean Runyan Associates. (2011). *The economic impact of travel on Texas*: 1990-2010p. Produced for Texas Tourism, Office of the Governor, Texas Economic Development and Tourism. <http://www.deanrunyan.com/doc_library/TXImp.pdf>

[249] Texas Port Association. (2011). *Port of Galveston.* <http://www.texasports.org/ports/ galveston/>

[250] Business Research and Economic Advisors. (2010). *The contribution of the North American cruise industry to the U.S. economy in 2009.* <http://www.cruising.org/sites/ default/files/pressroom/2009Economic Studies/EconStudy_Full_Report_2009.pdf> p. 56.

[251] McKenna, K. (2009). *Texas coastwide erosion response plan*: 2009 update. <http:// www.glo.texas.gov/what-we-do/caring-for-the-coast/_documents/coastal-erosion/ response-plans/coastwide-erosion-response-plan.pdf> p. 3.

[252] Texas Parks and Wildlife. (2001). *West Galveston Bay on the mend.* Texas Wetland News (July): 1. <http://www.tpwd.state.tx.us/publications/pwdpubs/media/pwd_br_r 0400_0003_07_01.pdf>

[253] Texas Parks and Wildlife. (2001). *West Galveston Bay on the mend.* Texas Wetland News (July): 1. <http://www.tpwd.state.tx.us/publications/pwdpubs/media/pwd_br_r 0400_0003_07_01.pdf>

[254] Texas Coastal Management Program. (2010). *Texas coastal and estuarine land conservation plan.* <http://coastalmanagement.noaa.gov/mystate/docs/celclplantx.pdf> p. 13.

[255] Texas Coastal Management Program. (2010). *Texas coastal and estuarine land conservation plan.* <http://coastalmanagement.noaa. gov/mystate/ docs/celclplantx.pdf> p. 13.

[256] Texas Forest Service. (2011). *Sunset Commission study: Transferring state forests to TPWD.* <http://www.sunset.state.tx. us/82ndreports/ tfs/studies/tpwd.pdf>

[257] Texas Parks and Wildlife Department. n.d. Texas state fish hatcheries. <http://www.tp wd.state.tx.us/fishboat/fish/management/hatcheries/>

[258] Texas State Data Center. n.d. Texas State Data Center. <http://txsdc.utsa.edu> (As cited in Jacob, J.S., R. Lopez, and H. Biggs. 2007. Anticipated local response to sea level rise along the Texas coast: A first approximation. <http://www.urban-nature.org/publications/documents/FinalDraft_AllCounties_Texas_SLR_Response_7_01_08.pdf> p. 11.)

[259] National Oceanic and Atmospheric Administration. (2011). National Climatic Data Center report, <http://www.ncdc.noaa.gov/img/climate/research/tornado/small/avgt5304.gif>

[260] Gibeaut, J. (2011). *Changes along the Texas Barrier-Island coast. Presented at the Gulf Coast Ecosystem Restoration Task Force meeting.* <http://www.epa.gov/gcertf/pdfs/830amtxgibeaut.pdf>

In: Gulf Coast Ecosystem Restoration Strategy ...
Editors: Bradley F. Brown and Cody B. Davis

ISBN: 978-1-61942-781-5
© 2012 Nova Science Publishers, Inc.

Chapter 2

AMERICA'S GULF COAST: A LONG TERM RECOVERY PLAN AFTER THE DEEPWATER HORIZON OIL SPILL*

Ray Mabus

> *"We need to understand that the environment and our lifestyle and our communities and the businesses that flourish down here must work together."*
> (Town Hall Participant, Houma, Louisiana)

The Deepwater Horizon oil spill was one of the worst man-made environmental disasters our country has ever experienced. The oil spill has dramatically affected the lives, jobs, and futures of millions of Gulf Coast residents. The Gulf of Mexico is a natural resource of vital importance which provides immeasurable benefits and services to citizens throughout the United States. The Gulf is also critical to nationwide commerce. Over time, the countless demands made on the region have critically impacted the entire Gulf environment. The most recent, and most damaging, of the impacts to the Gulf is Deepwater Horizon.

Today, the well is dead, oil no longer flows into the Gulf, and the attention of the media and the public is beginning to shift elsewhere. But as the president said on June 15, 2010, we will not forget what occurred, and we will not forget the promise made to the people of the Gulf Coast to help them restore their ecosystems and economy to health. The effects of the oil spill may reverberate in the region and across the country for years to come.

What happens in the Gulf of Mexico affects America. Nearly one third of the seafood harvested in the continental United States, as well as 30 percent of oil production and 13 percent of our natural gas production, come from the Gulf. The Gulf is an environmental treasure and central to the nation's economy. America needs the Gulf. America needs the Gulf to be clean. America needs the Gulf to be healthy. America needs the Gulf to be sustainable.

For the past several months, at the president's direction, I have examined the question of what comes next. What can our country do to make good on our commitment to the region? This report is the result of that examination. It is the result of many trips to the Gulf Coast and meetings and conversations with fishermen; health officials; environmental workers; nonprofits and local leaders; scientists; members of the business community; elected officials;

---

* This is an edited, reformatted and augmented version of the America's Gulf Coast, dated September 2010.

and thousands of Gulf Coast residents who expressed their opinions in meetings and town halls across all five states that touch the Gulf.

Throughout, I heard several common themes expressed over and over again: the plans put in place to restore the Gulf must come from the people of the Gulf; they must be based on science and good research; and the responsible parties must pay for the effects of the oil spill. Additionally, people of the Gulf Coast believe the impacts of years of economic and environmental damage resulting from coastal erosion and environmental neglect should be addressed by those who have benefitted from the Gulf's resources.

In a town hall in Ocean Springs, Mississippi, I heard the best expression of the plan that is needed: *"Recovery and sustainability for the Gulf depends on three critical resources: our people; our environment; and our commerce. We need a recovery plan that brings these aspects back into balance."*

This report will focus on these three central requirements for recovery in the Gulf Coast: the environment; the economy; and health and human services. These three areas are inextricably linked — what happens to the ecosystem of the Gulf affects its economy and the welfare of all Gulf Coast residents. As the Gulf ecosystem is rebuilt, economic activity will rise, jobs will be created, and the region's health will improve. The report will also discuss the role of nongovernmental organizations and community service in addressing each of the three issue areas.

During the town hall meetings, I heard one other theme: The need for dedicated recovery funding. As a gentleman told me in Theodore, Alabama: *"I've seen so many starts, but then, because there wasn't a dedicated funding source, nothing happened."*

A key recommendation of this report will be to call on Congress to dedicate a significant amount of any civil penalties obtained from parties responsible for the oil spill under the Clean Water Act to the recovery of the region that was damaged, and to those impacted by its effects. The report will outline a recommendation for establishment of a congressionally mandated governance structure to oversee and implement these and other sources of funding Congress may appropriate with the goal of a coordinated federal, state, and local long-term recovery strategy.

It is clear that as we move forward, there must be a seamless transition from response to recovery. In the five months since the Deepwater Horizon tragedy, many effective mechanisms have been put in place to respond to the environmental, economic, and health impacts of the spill. What has already been done cannot be lost, but rather must be built upon and expanded. To accomplish this, the report recommends leadership in overseeing the transition as well as the immediate establishment of a new Gulf Coast Ecosystem Restoration Task Force. This intergovernmental structure will be lean, flexible, adaptive, and able to rapidly incorporate the ideas of the communities it is designed to serve.

The task ahead is difficult and the actions recommended in this report will not resolve every problem faced but they can serve as building blocks for a better future in the Gulf. Our country has repeatedly faced big challenges, but it is exactly then that we display what is best about America. We never give up. We never give in. Instead, we band together to work toward a better future. Together, we will help make the Gulf of Mexico and the entire Gulf Coast whole again —for its citizens, and for all America.

Ray Mabus
Secretary of the Navy

AMERICA'S GULF COAST

I. Introduction

"This is a national issue to try to restore the Gulf. It's not just the Gulf States that are at stake here."
Town Hall Participant, Galveston, Texas

America's Gulf Coast is a national treasure. Its natural resources are an important economic engine for the entire United States; its waters sustain a diverse and vibrant ecosystem; and the Gulf's culture, natural beauty, and historic significance to the birth and growth of America are unique and priceless. Each year, millions of tourists visit the Gulf to vacation, swim, boat, fish, hunt, and birdwatch. Together, the Gulf's tourism and commercial and recreational fishing industries contribute tens of billions of dollars to the U.S. economy.[1] More than 90 percent of the nation's offshore crude oil and natural gas is produced in the Gulf,[2] and the federal treasury receives roughly $4.5 billion dollars every year from offshore leases and royalties.[3] And it is in the Gulf of Mexico that nearly one third of seafood production in the continental U.S. is harvested.[4] America needs a healthy and resilient Gulf Coast, one that can support the diverse economies, communities, and cultures of the region.

In the wake of the Deepwater Horizon disaster, what has become clear is that the economy of the Gulf Coast and the health of its citizens are inextricably linked to the health of the Gulf's waters and shoreline. A significant portion of the jobs in the region are connected to companies and small businesses involved in tourism; commercial and recreational fishing; and related services. Because of the spill, these industries have lost income as confidence in the safety of the Gulf's beaches and seafood has fallen. Many tourists have chosen to vacation elsewhere and the Gulf's beautiful coastline has been marred in the public eye by the constant television presence of orange oil booms and tar balls on the beaches.

The spill has exacerbated the effects of a multitude of storms and years of environmental decline. Building on these earlier problems, the oil spill created economic uncertainty for hundreds of thousands of Americans, and its social consequences and behavioral health effects have stretched the capacity of nonprofit organizations and local governments. Moreover, local, state, and tribal governments face the prospect of diminished resources to deal with these issues due to lower tax revenues from lost economic activity and diminished property values.

The Gulf Coast recovery process must address a range of effects of the oil spill. This report identifies these effects and provides recommendations for comprehensive Gulf Coast recovery. The report is not structured to distinguish between needs that may be addressed through legal claims brought by individuals, businesses, and governmental entities against the parties responsible for the spill and those that will need to be addressed through a combination of existing federal, state, and local programs and proposals for new congressional action.

For example, the Oil Pollution Act of 1990 (OPA) provides for recovery by designated state and federal trustee agencies of natural resource damages to address injuries to natural resources resulting from oil spills. These funds must be spent in accordance with OPA's directives. Therefore, damages recovered under this provision for injuries caused by the Deepwater Horizon spill may address many, but not all, of the natural resource needs identified in this report. Other existing funding sources may be available to aid in this effort, in accordance with governing statutes.[5]

Integral to the recovery also must be the empowerment of the residents of the Gulf Coast. We must establish conditions that will allow the people of the Gulf to build upon the recovery process and create a more resilient Gulf Coast for the future.

II. An Unprecedented Crisis and Response

On April 20, 2010, the Deepwater Horizon oil rig exploded in the Gulf of Mexico 41 miles southeast of the Louisiana coast. The explosion killed 11 people, and released a torrent of oil and gas into the Gulf that spewed unrelentingly for almost three months, until the well was successfully capped in a coordinated effort on July 15, 2010.[6]

By the time the well was capped, it is estimated that the Deepwater Horizon had released more than 4.9 million barrels of oil into the Gulf,[7] resulting in significant impacts on the Gulf's communities, ecosystem, and economic activity. It is one of the worst man-made environmental disasters in American history. The spill caused the closure of 88,522 square miles of federal waters to fishing, and affected hundreds of miles of shoreline, bayous, and bays.[8] While the full long-term effects of the spill will take time to determine, the negative environmental, economic, and social impact on the Gulf's habitats and communities may persist for years.

In response to the spill, President Obama called for immediate and coordinated federal, state, local, and industry action. At its peak, efforts to stem the spill and combat its effects included more than 47,000 personnel; 7,000 vessels; 120 aircraft; and the participation of scores of federal, state, and local agencies.[9] The dedicated efforts of the thousands of men and women involved mitigated the worst impacts of oil on shorelines across the Gulf. The response led to the deployment of nearly four million feet of oil boom,[10] and the recovery of approximately one million barrels of oil.[11]

The federal response also took steps to care for people impacted by the spill. At the president's request, BP committed $20 billion to a trust fund designed to provide compensation for damages incurred by individuals and businesses, as well as for certain government claims. Responsibility for adjudicating individual and business claims against BP to be paid out of this fund was turned over to an independent claims facility run by Kenneth Feinberg, who on August 23, 2010 opened the Gulf Coast Claims Facility to manage the process. Additionally, the National Incident Command (N IC) established the Integrated Solutions Team (IST) to develop and coordinate support services for individuals and small businesses impacted by the spill's effects.

Agencies across the federal government have reached out to local communities. In particular, an interagency Economic Solutions Team (EST) composed of representatives from the departments of Agriculture, Commerce, Homeland Security, Labor, the Small Business Administration, and other agencies, have worked with local communities to listen to their concerns. The EST is working to help create local solutions to the problems facing Gulf residents. This has included sending 21 assessment and evaluation teams to counties, parishes, and communities across the Gulf to help them begin facilitating economic recovery.

Other departments and agencies, including the Department of Health and Human Services, (HHS), the Department of Labor, the Environmental Protection Agency (EPA), The Corporation for National and Community Service (CNCS), as well as state and local governments and nonprofits, are responding to the health and human services needs of the Gulf Coast.

The ecosystem protection and restoration effort is also already underway. Several federal agencies, including the Department of Agriculture, the National Oceanic and Atmospheric Administration (NOAA), the EPA, the U.S. Fish and Wildlife Service (USFWS), the U.S. Geological Survey (USGS), and the National Park Service, in coordination with state environmental agencies and nonprofit groups, have supported response operations with the best available science and years of experience working in the Gulf.

In addition, a Natural Resource Damage Assessment (NRDA) is underway. The function of this process is to assess environmental harm in order to restore natural resources and natural resource services injured by the spill. The USFWS, on behalf the Department of the Interior serves as the Federal Lead Administrative Trustee on the Natural Resource Damage Assessment Trustee Council.

The government's response to this latest disaster is guided primarily by the Clean Water Act of 1977 (CWA), the Oil Pollution Act of 1990 (OPA), and related regulations. The CWA has a primary goal of restoration and maintenance of the Nation's waters and provides several causes of action enforceable by the United States in order to promote these goals. These include provisions for the recovery of civil (and criminal) penalties. The CWA civil penalty provision associated with oil spills provides that penalties recovered under the Act must be deposited into the Oil Spill Liability Trust Fund. The Fund, in turn, is designed, among other things, to ensure that there are available funds for clean-up, response, and restoration efforts for future oil spills. The Fund is available to pay compensation for removal costs and damages if a responsible party does not do so and to pay compensation in excess of the responsible parties' liability. OPA, enacted in the wake of the 1989 *Exxon Valdez* oil spill, makes parties responsible for oil spills liable to pay the costs to remove the oil as well as to pay certain damages resulting from the spill, subject to liability caps. OPA damages include several categories of economic damages as well as damages associated with injuries to natural resources. OPA also amended sections of the CWA to mandate new contingency planning and response preparedness responsibilities for both industry and the federal government. Finally, as mentioned above, OPA provides for an assessment of natural resource damages from oil spills and mandates that, subject to certain statutory limitations, responsible parties pay the cost of these damages.

The actions taken thus far are the first steps towards recovery. In his Oval Office address on June 15, 2010, the president committed the nation to help the Gulf Coast and its people recover from this tragedy, to make a commitment to the Gulf that would go beyond the crisis of the moment. To fulfill that pledge, the report recommends a set of additional actions.

III. A Way Forward

This report recommends two parallel and complementary efforts to ensure a seamless recovery and restoration effort in the Gulf Coast.

In order to help address the harm inflicted upon the region, dedicated funds are absolutely essential. This report recommends that the president urge Congress to dedicate a significant amount of any civil penalties recovered under the Clean Water Act from responsible parties toward assisting the region where the damage from the spill occurred. The report also recommends that Congress establish a Gulf Coast Recovery Council to coordinate the federal, state, local, and tribal actions that will be taken, funded in part with financial support from CWA civil penalties, to restore the Gulf Coast. The Gulf Coast Recovery Council would also coordinate with and provide support to the trustee agencies charged with seeking natural resource damages and conducting restoration through the NRDA process.

It is critical that there be a seamless transition from response and clean up to recovery and restoration in parallel with congressional action. To accomplish this, the report recommends the immediate establishment of a federal lead for Gulf recovery and the creation of a Gulf Coast Ecosystem Restoration Task Force. The function of the Gulf Coast Ecosystem Restoration Task Force will be to coordinate the recovery of the region's ecosystem. If Congress acts to establish the proposed Gulf Coast Recovery Council, the Gulf Coast Ecosystem Restoration Task Force could be modified or dissolved, with its duties and coordinating bodies subsumed by the Gulf Coast Recovery Council as directed by Congress.

It is also recommended that the Departments of Health and Human Services, Commerce, Agriculture, and Labor, in conjunction with the Small Business Administration (SBA), the Internal Revenue Service (IRS), the Corporation for National Community Service (CNCS), and other agencies, continue their important recovery work in close coordination with state and local governments and organizations.

In summary, this plan encompasses five topic areas critical to the long-term recovery of the Gulf region:

- Proposal to Congress to Dedicate Clean Water Act Civil Penalties to the Gulf Coast
- Long-Term Ecosystem Restoration
- Health and Human Services Recovery
- Economic Recovery
- Nonprofit Sector Recovery

IV. Congressional Action

"We need a strong funding mechanism to support the many aspects of Gulf restoration."

(Town Hall Participant, St. Petersburg, Florida)

Restoring the Gulf Coast, assisting in its full recovery, and reversing the effects of the Deepwater Horizon oil spill on its economy, environment, and public health are not just local or regional priorities, they are national issues. It is in the Gulf where much of the nation's oil and seafood are produced, it is in the Gulf where the damage from the spill occurred, and it is to the Gulf where recovery funds must be dedicated. This section outlines a proposal for Congress to create a new Gulf Coast Recovery Council that would be funded in part by civil penalties collected under CWA, and which would work to facilitate environmental restoration and economic recovery, and attend to health issues arising from the spill.

A. Dedication of Clean Water Act Civil Penalties to the Gulf Coast Recovery Effort

It is recommended that the president urge Congress to pass legislation that would dedicate a significant amount of any civil penalties recovered under the Clean Water Act from parties responsible for the Deepwater Horizon oil spill to those directly impacted by that spill. As previously discussed, the Clean Water Act directs that civil penalties associated with oil spills must be deposited into the Oil Spill Liability Trust Fund. The Fund is designed, among other things, to ensure that there are available funds for cleanup and response efforts for future oil spills. Without additional legislation, these funds cannot be deposited into a Gulf Coast Recovery Fund. Deepwater Horizon was more than an order of magnitude greater than any other oil spill the nation has faced to date. Therefore, a new mechanism to help fund overall Gulf restoration and recovery operations is necessary.

Specifically it is recommended that the president propose to Congress that legislation be passed that would allow:

- A significant amount of any civil penalties recovered under the Clean Water Act from the Deepwater Horizon spill be deposited into a Gulf Coast Recovery Fund managed by a Gulf Coast Recovery Council. These funds would be used to address those critical recovery needs that may fall outside the scope of the OPA.
- The establishment of a Gulf Coast Recovery Council to lead to long-term ecosystem, economic, and health recovery in the Gulf.
- A portion of any Clean Water Act civil penalties be directed to the Gulf states (Alabama, Florida, Louisiana, Mississippi, and Texas) to enable them to jumpstart their own recovery efforts.
- The remaining amount of penalties be deposited in the Oil Spill Liability Trust Fund in accordance with existing law. This would allow for resources to be available to respond to future spills.

B. Additional Request to Congress

In addition, this report recommends that the president urge Congress to act on the legislative priorities that the Administration has already outlined in the context of a supplemental package that was submitted on May 12, 2010 and included both funding and authorizing language intended to facilitate a response that would be expedient, deliver speedy assistance to people affected by this spill, and strengthen and update the oil spill response system.

Many of the provisions included in the package were included in legislation (H.R. 4899) which was enacted in late July, 2010. (P.L. 111-212). However, H.R. 4899 excluded Administration- proposed provisions that increase the availability and flexibility of:

- Unemployment Insurance: The Administration requested a new program of unemployment assistance, modeled after the Disaster Unemployment Assistance program, to help workers who are unemployed as a result of a spill of national significance.
- Employment Assistance: The Administration's request proposed funding and additional flexibility to temporarily expand the service capacity of Workforce Investment Act training and employment programs at the state and local levels through the National Emergency Grant (NEG) program.
- Liability Limits: In addition, the Deepwater Horizon tragedy has shown that the current liability regime under OPA needs to be updated. Congress should remove the liability cap for damages associated with offshore drilling activity and increase the liability caps for other activities that could result in a spill.

There may be other cases where more flexibility would be helpful, and this report recommends working with Congress to review the May request and identify needs that have become apparent since then.

C. Responsibilities of the Gulf Coast Recovery Council

This report recommends that Congress establish a Gulf Coast Recovery Council that should focus on improving the economy and public health of the Gulf Coast, and on ecosystem restoration not dealt with under NRDA. These three areas are inextricably linked to the successful recovery of the region.

The Council should include representatives from the following departments and agencies: Agriculture, Commerce, Corporation for National Community Service, Energy, Environmental Protection Agency, Health and Human Services, Housing and Urban Development, Interior, Justice, Labor, the Small Business Administration, Treasury, and the U.S. Army Civil Works/Corps of Engineers. The Council should also include representatives from states and from federally-recognized Gulf tribal organizations, who would be appointed by the president. These entities would have the opportunity to recommend possible candidates to the president.

While it would be a separate body, the Council should support, closely coordinate with, and possibly share overlapping membership with the Natural Resource Damage (NRD) Trustee Council in order to ensure coordinated ecosystem restoration actions across the region without preempting the NRDA decisions of the Trustee Council.

Using any CWA civil penalties that Congress dedicates to Gulf Coast recovery and any other available funds, a congressionally mandated Council should support projects that will restore the ecosystem to health or make it more resilient. The damage from the oil spill was not, however, just environmental. The spill hurt the Gulf's ability to market its seafood and tourism; it harmed small businesses across the region; and it negatively affected the physical and behavioral health of thousands of Gulf residents. Therefore, funds should also be used to strengthen the health of the regions residents and make the Gulf's economy adaptable to future economic dislocations by expanding markets, diversifying products, and strengthening the workforce.

D. States and Local Communities Must Lead Their Own Recovery

State and local governments must be allowed to lead their own recovery, and should be provided resources and a prominent role in Gulf Coast recovery. State leadership will be facilitated by the proposal that Congress dedicate an amount of any Clean Water Act civil penalties recovered from the Deepwater Horizon oil spill directly to state governments.

- To ensure adequate state representation, in addition to a presidentially-appointed federal chair, this report recommends that the leadership of the Gulf Coast Recovery Council should also include a presidentially-appointed state co-chair. The governors of the Gulf states would have the opportunity to recommend possible candidates to the president.
- Local governments and citizen stakeholders must play a critical role in the Gulf Coast Recovery Council. The Gulf Coast Recovery Council must be allocated the resources to maintain adequate intergovernmental outreach, public affairs, and communications personnel to be effective while remaining lean and efficient.

E. The Gulf Coast Can Recover Stronger and Better

Through the recovery process, the Deepwater Horizon oil spill presents a unique opportunity to go further and make the Gulf healthier and more resilient than it was before the spill. Because the recovery process, personnel, and equipment will already be in place as a result of the NRDA process, dedicated penalty funds can provide the flexibility to tackle issues related to, but not specifically caused by, the spill. While these funds may not fully address all of the challenges facing this region and this ecosystem, they will allow for a level of effort and focus that is unprecedented. A congressionally chartered Gulf Coast Recovery Council would oversee management of these tasks during the recovery effort.

F. Participation and Transparency are Critical

The Gulf Coast Recovery Council should seek out participation from a wide array of stakeholders, involve the private and nonprofit sectors, be transparent and accountable to taxpayers, and maintain the confidence of the public.

G. Scientific Coordination and Audit Functions are Crucial

The Gulf Coast Recovery Council should work with existing federal and state advisory committees, as appropriate, to ensure that relevant scientific and technical knowledge underpins recovery planning and decision making, and that research, monitoring, and assessment efforts are organized. The Council should also provide oversight and accountability into Gulf of Mexico recovery efforts by developing quantifiable performance measures that can be used to track progress towards recovery goals.

V. Other Potential Funding Sources

This report makes recommendations on programs and efforts that may require additional funding beyond what can be addressed through existing legal mechanisms. Without some combination of Clean Water Act civil penalties or successful requests to responsible parties, it is possible that many funding priorities laid out in this report will not be immediately realized. That is why it is critical both that Congress act and that the parties responsible for the spill play their part in the recovery of the Gulf Coast.

A. Private Action

An idea that has arisen on numerous occasions is the need for an independently funded, nongovernmental recovery effort. This would be an opportunity for local and regional leadership to tackle issues not funded through the Oil Pollution Act or other existing sources. Individuals and companies with a significant interest in the economic and environmental health of the Gulf Coast have a role to play in recovery. Private action is an important component of Gulf Coast recovery and could serve as a long-term complement to the federal government response of seeking congressional action.

B. Requests to Responsible Parties

To date, some federal agencies have secured from BP initial funding to support immediate response and recovery needs beyond the ecosystem. HHS's Substance Abuse and Mental Health Services Administration (SAM HSA) successfully coordinated with state agencies to gain $52 million in behavioral health services and surveillance funds from BP; and HHS's National Institutes of Health (NIH) received $10 million in public health research funding. Going forward, the Administration, working with state and local governments, should develop a unified additional request to the responsible parties for the oil spill for economic recovery and public health needs.

- In coordination with state public health and behavioral health departments, HHS and other agencies should continue their efforts to identify health and human services funding needs that should be met by BP and other responsible parties. Some examples of potential needs include funding for surveillance, behavioral health services, and research.
- Building on work already being done by the states, the Department of Commerce and EPA should lead an effort in partnership with state fish and wildlife agencies to develop a coordinated plan and funding requirements for seafood certification and marketing.
- In partnership with state and major municipal governments, the administration should lead efforts to develop a coordinated plan that responsible parties should fund to promote tourism.

C. Accelerating the Flow of Coastal Impact Assistance Program (CIAP) Funds

CIAP was created by Congress in 2005 to allow coastal states to provide for ecosystem restoration, and to help mitigate the impacts of oil and gas production on the Outer Continental Shelf. Under CIAP, the Secretary of the Interior is authorized to distribute $250 million for each of the fiscal years 2007 through 2010 to oil and gas producing states and coastal political subdivisions.

Following Deepwater Horizon, the Department of the Interior is working to implement necessary changes to CIAP to speed the flow of remaining funds to the four eligible Gulf

States (Alabama, Louisiana, Mississippi, and Texas). Accelerating CIAP could send up to $598 million in additional funding for ecosystem projects to the Gulf.

VI. Transition from Response to Recovery

In order to ensure that there is continuity between the response efforts led by the National Incident Command (NIC) and proposed recovery efforts, this report recommends that a structure be established by administrative action through existing authorities. (This should be in addition to the ongoing work being conducted by the Natural Resource Trustees through the NRDA process). Administrative action should formalize the environmental work already underway by establishing an intergovernmental Gulf Coast Ecosystem Restoration Task Force.

The departments of Commerce, Agriculture, and Labor, along with the SBA, IRS, HHS, and CNCS, should continue their respective economic and public health efforts with states and local governments, and nonprofit organizations. If Congress dedicates any obtained Clean Water Act civil penalties to the Gulf Coast Recovery Fund and establishes the Gulf Coast Recovery Council, this Gulf Coast Ecosystem Restoration Task Force may be modified or dissolved, with its duties and coordinating bodies subsumed by the Recovery Council.

It is also essential that there be a single presidentially-appointed leader to oversee the transition from cleanup and response to recovery and restoration. This individual could serve as the chair of the Gulf Coast Ecosystem Restoration Task Force and as the single point of contact within the administration to help coordinate federal restoration and outreach efforts.

A. Gulf Coast Ecosystem Restoration Task Force

The images of oil drenched pelicans and tar soaked beaches may no longer headline the news, but the environmental impacts of the Deepwater Horizon oil spill are uncertain. The impacts of the spill both exacerbated ongoing environmental issues and caused new injuries. Federal and state scientists and independent researchers are working to understand the harm to natural resources caused by the spill.[12] From an ecological standpoint, the impacts on the water column, fisheries, habitats, and diverse species of the Gulf are of great concern.

Plans to restore the Gulf must come from the people of the Gulf. To effectively address the damage caused by the Deepwater Horizon oil spill, address the long-standing ecological decline, and begin moving toward a more resilient ecosystem on the Gulf Coast, the federal government must work with state and local governments, non-profit organizations, community groups, scientists, academics, the private sector, and the public.

In order to facilitate the implementation of programs and projects – those needed to repair harm from the spill, as well as those envisioned before the spill –this report recommends a unifying structure be created to facilitate efforts between disparate federal, state, and local stakeholders.

A lean federal-state Gulf Coast Ecosystem Restoration Task Force should immediately be established to support the Natural Resource Damage Assessment (NRDA) process and coordinate non-N RDA ecosystem funds, programs, and projects. While the task force would not direct the actions of other federal agencies, it would serve in a critical advisory capacity to ensure that Gulf restoration efforts are coordinated, collaborative, and effective.

Membership

The Gulf Coast Ecosystem Restoration Task Force would include federal and state representatives, as well as representatives from federally recognized tribes. Federal membership on the task force could include the departments of Agriculture, Commerce, Interior, Justice, and Transportation, along with the Environmental Protection Agency, U.S. Army Civil Works/Corps of Engineers, the Council on Environmental Quality, the Office of Science and Technology Policy, the Domestic Policy Council and the Office of Management and Budget. The departments of Health and Human Services, Homeland Security, Labor, and the Small Business Administration and other agencies could serve in an advisory role to offer guidance on issue-specific topics.

Each Gulf state could have one representative on the Gulf Coast Ecosystem Restoration Task Force, nominated by each Gulf state governor and appointed by the president. In many cases, individuals who would serve on the task force from the state agencies are likely to be the same as those serving as their state's representative on the Deepwater Horizon oil spill NRDA Trustee Council. Such overlap in membership would be encouraged to help ensure coordination with NRDA restoration activities. It is important that the task force reach out to federally recognized tribes in the region to engage them in undertaking its responsibilities, in collaboration with them, and determine an appropriate structure for tribal participation.

Leadership

The Gulf Coast Ecosystem Restoration Task Force would be chaired by a federal representative. One state representative, jointly selected by the state representatives on the Gulf Coast Ecosystem Restoration Task Force, would serve as vice-chair.

Responsibilities of the Gulf Coast Ecosystem Restoration Task Force

The role of the Gulf Coast Ecosystem Restoration Task Force would be to draw from the expertise and implementation mechanisms of the various existing programs and authorities and suggest ways to better align and coordinate those efforts across federal agencies and state governments. Key responsibilities of the task force should include:

- Facilitate Project Implementation: The Gulf Coast Ecosystem Restoration Task Force should develop ways to make restoration project review and implementation more effective in developing and implementing science based projects. It would be a priority for the task force to identify and work under existing agency authorities and programs to resolve barriers to implementation for restoration projects identified by its members, as well as to address needs raised by the Deepwater Horizon oil spill NRDA Trustee Council.
- Program and Authority Coordination: The Gulf Coast Ecosystem Restoration Task Force should work with the federal and state agency members on improving the alignment of each agency's relevant programs and authority.
- Coordination with Other Agencies: The Gulf Coast Ecosystem Restoration Task Force also would be expected to coordinate with HHS on health and human services issues and with Commerce, Labor, SBA, Treasury and other federal agencies on ways to maximize the economic impact of ecosystem restoration projects.

- Budget Coordination: The Gulf Coast Ecosystem Restoration Task Force should coordinate with OMB in developing annual budgetary guidance on Gulf of Mexico priorities for all member agencies.
- Oversight and Accountability: The Gulf Coast Ecosystem Restoration Task Force should work with agencies to develop performance measures and prepare a biennial report card on progress.

Stakeholder Outreach

The Gulf Coast Ecosystem Restoration Task Force should conduct its operations in such a way as to ensure that stakeholder groups have the ability to express their views. It is important to create formal mechanisms for public outreach and stakeholder engagement and information sharing. All relevant options should be reviewed and evaluated to ensure formal, comprehensive public engagement of the Gulf community throughout restoration efforts.

Science Coordination and Research

The Gulf Coast Ecosystem Restoration Task Force should work with existing federal and state advisory committees and the NRDA Trustee Council, as appropriate, to ensure that relevant scientific and technical knowledge underpins restoration planning and decision making, and that research, monitoring, and assessment efforts are organized, in consultation with the Office of Science and Technology Policy.

The Gulf Coast Ecosystem Restoration Task Force should also coordinate with colleges and universities and other related efforts across the Gulf Coast, such as the National Oceans Council, Hypoxia Task Force, and Gulf of Mexico Alliance, who are studying the unique ecosystems and economic conditions of the region. These institutions have years of expertise working on the Gulf and are trusted by Gulf residents. The task force should also coordinate with other research institutions and among federal agencies. In addition, it should look to enable innovative and sustainable solutions for restoration of coastal ecosystems, wetlands and estuaries, whether through academic institutions, private enterprise, non-profits or governmental entities.

B. Health and Human Services Efforts

HHS has been coordinating public health efforts with federal agencies, states, and local communities. This report recommends that HHS maintain this function and serve in an advisory capacity to the Gulf Coast Ecosystem Restoration Task Force.

In the context of a complex disaster like the oil spill, it is critically important to ensure that rigorous surveillance inform the science and research agenda and recovery actions, including behavioral health prevention and treatment, the delivery of medical services, and access to human services. Likewise, ensuring seafood safety not only protects human health, but promotes economic recovery across the region. The health response on the Gulf Coast must be based on empirical evidence made possible by targeted research.

Much has been learned from prior efforts about the critical need to promote cooperation and coordination between federal, state, and local entities in order to align, refine, and adapt recovery plans and projects. The evolving nature of the oil spill's effects requires a particularly active collaborative process governed by community needs and carried out through local, state, and federal government, as well as through public-private partnerships.

Surveillance

Appropriate surveillance is essential for detecting harmful health effects, understanding behavioral health needs, assessing the impact of behavioral health interventions, and monitoring human services needs related to the spill. Surveillance is taking place across the Gulf Coast, and needs assessments have begun. However, in order to allow for long-term needs assessment and planning activities, all of the Gulf States need both technical assistance and additional funding to expand their current capacity for surveillance and data collection.

Behavioral Health

Over 20.5 million people live on the Gulf Coast and are dependent on the area for their livelihoods. Behavioral health consequences will be especially prevalent in those whose livelihoods depend on industries affected by the spill. State behavioral health providers have indicated that some in the affected population are already exhibiting early signs of developing substance abuse and dependence, psychiatric disorders, suicidal risk, and family breakdown, including domestic violence and child abuse. Responding to these needs now will positively address the behavioral health consequences of the spill.

Medical Services

Health-related claims will need assured access to the claims process as long-term health effects are identified. Dedicated and reliable funding must be put in place to guarantee access to care for those with oil-related health effects. Reimbursement for emergency departments, definitive care facilities, and specialty care providers that treat oil spill-related health effects must be reliable, efficient, and timely.

Human Services

A reduction in resources caused by the Deepwater Horizon oil spill will likely increase demand for human and social services designed to assist impacted populations. Job training, child care assistance, early childhood education programs, nutrition assistance, domestic violence response, senior services, and cash assistance programs may all see increased demand. Federal efforts to address these concerns must consist of three components: monitoring; making best use of existing programs and funding; and creating new programs as appropriate.

Seafood Safety

As mentioned throughout this report, restoration of confidence in seafood safety will be a longterm effort on the Gulf Coast. HHS, NOAA, and EPA are committed to this effort and have identified several recovery activities such as long-term surveillance and inspection requirements to ensure that seafood harvested from Gulf waters is safe. It is essential that state and federal laboratories be able to receive, analyze, confirm, and report the presence of oil, dispersants, and other toxic agents in seafood, clinical, and environmental samples.

Research

Although there are many studies on research into spill response technologies and on the effects of oil spills on the environment, there are surprisingly few studies on the long-term effects of oil spills on human health. Most studies to date have focused on short-term effects such as skin, eye, nose and throat irritations and not on the long-term health impacts of a spill.

It will be necessary to continue identifying and supporting research efforts to address community health concerns and fill gaps in knowledge about the health consequences of cleanup activities surrounding the oil spill. The research efforts span studies of seafood safety, response workers, and volunteer populations as well as studies of the general population and at-risk groups, including children and pregnant women. This research should continue to be funded.

C. Economic Recovery

It is critical that the economic response reflect local priorities, incorporate local input, and build from local visions for economic recovery. The spill occurred at a moment when communities across the region were actively engaged in planning new economic development. Those visions and plans cannot be lost. They must form the basis for federal, state, local, and private economic development efforts.

The oil spill presents a unique set of economic challenges for the Gulf region. This is a region where many workers are underemployed or work in jobs that pay below the national average, or have little or no safety net. The impact is grave on such populations that are not as resilient to downturns. As the effort progresses it is essential that recovery funds are directed into local economies, which will result in increased employment opportunities and heightened local economic activity.

The departments of Commerce, Agriculture, and Labor, in conjunction with the SBA and the IRS, should continue their economic recovery work in close coordination with other federal agencies and with state and local governments and organizations. These agencies should also coordinate closely with the Gulf Coast Ecosystem Restoration Task Force, in order to optimize the economic impact of ecosystem recovery and better leverage public and private sector investments in the region. This includes:

- Maximizing participation of the local labor force in contracting and job skills training and providing opportunities for disadvantaged and underemployed members of the population.
- Utilizing recovery projects to support economic diversification and communities that have been historically underserved and underrepresented.
- Working to fund green recovery projects and promote clean energy market development.
- Providing technical assistance to communities, small businesses, and individuals in navigating the recovery process.

Through conversations and meetings with fishermen, small business owners, local leaders, elected officials, and thousands of people who participated in town halls and community economic assessments, several recommendations emerged that should be considered by the federal and state response.

Compensation and Income Support for Affected Individuals and Businesses

For people across the Gulf, obtaining compensation for lost income and economic damages through the claims process is a critical short-term measure necessary to building economic health. Across the Gulf Coast, communities have expressed significant concern about the efficiency of the claims system and about their own capacity to use the process. Specifically, some communities have requested technical assistance to ensure that individuals and small businesses are able to file claims. The federal government has made significant efforts to ensure that those eligible for compensation have access to the Gulf Coast Claims Facility. These efforts will continue.

In addition, from the earliest days of the spill, the federal government has made it a top priority to ensure access to support services. The federal government, along with state and local governments and organizations, should:

- Continue efforts to integrate services and facilitate access to government programs. Integration of services for disaster victims should remain a priority for the administration. It is particularly important that individuals unemployed as a result of the oil spill be effectively connected to safety net programs such as unemployment compensation, Temporary Assistance for Needy Families, and the Supplemental Nutrition Assistance Program (SNAP formerly Food Stamps).
- Provide technical assistance with claims submissions.
- Work to address the financial education and financial access needs of affected individuals, including helping claims recipients manage their payments. The federal government and its partners should also work to address the specialized credit and counseling support needs of small businesses in the region.
- Offer resources for tax guidance to those affected by the spill or receiving compensation from the GCCF. The IRS has taken a number of steps to support taxpayers impacted by the spill such as assisting with questions and issues through their website, establishing a dedicated phone line, and conducting outreach programs.
- Increase federal efforts to drive local hiring in response and recovery funds.

Economic Assessment, Technical Assistance and Capacity Building for Affected Communities

To develop effective recovery plans, local communities should assess the scope of local damage and identify a desired path to recovery. However, communities often lack the capacity to do this effectively, particularly in the wake of a disaster, when limited resources and human capital are already stretched thin.

Meetings and conversations with communities and local governments point to multiple economic challenges associated with insufficient local capacity. To mitigate local challenges and provide additional capacity to disaster-impacted communities the federal government should support state and local government organizations in their work to:

- Build economic development strategies around community needs, and take particular efforts to address the needs of disadvantaged, underserved, and resource constrained communities.
- Publicize and distribute an economic toolkit to help communities understand and access federal programs that may be relevant across different phases of recovery. (The Tool Kit has been included as an appendix to this report).
- Develop Gulf Coast Regional Economic Assets and Economic Clusters. Economic cluster strategies may include such efforts as establishing regional organizations to run business incubators, or assisting new companies in accessing early capital.

Rebuilding Public Support for Affected Industries

Communities in the Gulf are already reporting that negative perceptions of Gulf waters, beaches, and seafood industry are having a harmful effect on tourism and recreational activities. While communities remain concerned about seafood and water-safety issues; the cleanliness of beaches; and the need for continued testing and monitoring; they are also looking for ways to balance these safety concerns with efforts to market core industries and restore the image of the Gulf as a desirable place to live, visit, vacation, and do business.

To date, there have been a number of significant efforts to bolster the image of the Gulf and its core industries. This effort must continue. Recommendations for the federal government, along with state and local governments and organizations, include:

- Providing grants for seafood marketing: NOAA is already working with the Gulf States on plans to restore national confidence in Gulf fishery products. These efforts should continue. For example, in an effort to show its support for the Gulf, the Department of Defense will continue pursuing opportunities to procure more Gulf seafood.
- Fishery disaster assistance: The state of fisheries and fishery dependent businesses should continue to be monitored in order to identify and address potential challenges as they evolve.
- Industry priorities: Tourism and marketing should remain a priority if future funds become available.
- Contractor flexibility: Legislation that would allow greater flexibility for federal contractors to give priority to organizations, firms, and individuals residing in areas affected by the spill for contracts directly related to the response and recovery efforts.

Targeted Small Business Support

Communities report that small businesses are challenged by the uncertainty of current circumstances. In particular, concerns about the future are contributing to a general wariness about whether to take on new loans and employees and invest in future business opportunities. Local banks report difficulty extending operating credit to businesses for payroll and working expenses and the economic uncertainty is dissuading investment from outside the region. Many of these small and locally owned businesses were recovering but on tenuous ground following the hurricane seasons of 2004 and 2005 and the national economic crisis further destabilized many of these businesses.

Small businesses are critical to the long-term health and economic recovery of the Gulf Coast. They are an economic engine for local economies and one of the first catalysts of economic recovery. Support to small businesses should include:

- On-site small business counseling: Across the five Gulf States, SBA has established five Small Business Development Centers, 15 Women's Business Centers, and 25 SCORE (small business mentoring and training) chapters, with a total of more than 950 counselors operating on the ground.
- Small business federal procurement opportunities: SBA should work with the General Services Administration and other federal agencies to identify federal procurement opportunities for small businesses in the five states.

- Community development block grant funding: Gulf Coast states can make applications to HUD for competitive Section 108 Loans now for small business assistance, which include: working capital loans, real estate acquisition, development, expansion, or equipment purchase loans.

Long-Term Economic Diversification and Resilience

While the immediate efforts in response to the spill have centered on compensation, assessment, marketing, and other near term priorities, a central component of the economic vision for the Gulf Coast involves diversifying the economic drivers that create jobs and increase wealth in the region. Communities across the Gulf Coast have almost uniformly articulated a strong desire to diversify, often noting that the spill led them to the conclusion that their economies are too narrowly concentrated.

The challenge in this area is to balance short-term spill responses with pre-existing long-term goals and economic challenges faced by many communities in this region. The vision of a healthy, diverse, 21st century economy has great potential to come to fruition in several key areas that are vital to the Gulf Coast's long-term success. Ultimately, the decision of whether and how to diversify an economy must be the decision of and driven by the local community. Efforts to diversify should focus on:

- Investing in export potential: Directing funds toward businesses that take advantage of new opportunities to sell in foreign markets will present new sources of revenue for Gulf Coast companies.
- Supporting the development of a 21st century workforce: Job training efforts will enable workers to enter new and potentially lucrative careers in new technologies and industries.
- Promoting of a clean energy economy: Residents of the Gulf Coast have expressed a desire to transition to a clean-energy economy. A number of efforts from both the public and private sector have helped prime the pump for additional investments in clean energy. For instance, the US Department of Energy recently awarded the Southeast Energy Efficiency Alliance $20 million to implement energy efficiency programs in 13 Southern cities including New Orleans. These programs are underway and are scheduled to increase their scale over a period of two to three years. The Department of Energy's American Recovery and Reinvestment Act (ARRA) projects have also provided the Gulf Coast with the initial seed money necessary to make sure that green technology and green jobs are an integral part of the 21st century Gulf economy.

- Increasing federal efforts to drive local hiring in response and recovery funds: Ensuring that resources are used to bolster employment opportunities for Gulf Coast residents will help drive economic recovery.

D. Nonprofit support

Nonprofit organizations play a critical role in augmenting the support that federal, state and local governments provide to the residents in the Gulf Coast, as do philanthropic, faith-based and business interests. As such, their success will be an integral part of the recovery effort in the region. In order to help ensure the vitality of these important groups, it is recommended that nonprofits consider taking the following actions:

- Enhancing coordination: Improved methods for sharing information among nonprofits; local communities; and state and federal governments should be developed. In addition, efforts should be made to engage community and faith-based organizations.
- Increasing public awareness of the needs related to the oil spill: Advertising campaigns should be conducted to inform the public of opportunities to help with the recovery. These campaigns will be aimed at increasing philanthropic and volunteer participation in recovery efforts.
- Leverage Volunteers and Community Service Assets: In addition to soliciting support from the federal government and the philanthropic community, nonprofits should be encouraged to engage volunteers and other organizations dedicated to providing free services.
- Conducting additional needs assessments: Additional studies should be conducted to determine how nonprofits can help with recovery efforts.
- Enhance the fundraising capacity of nonprofits: Resources should be provided to help nonprofits better identify and leverage new and existing funding sources (for example, providing technical assistance related to grant writing). Additional dedicated funding resources to meet specific needs of affected communities and residents may be needed.

VII. Conclusion

This report provides recommendations for funding the recovery effort in the wake of the spill, as well as mechanisms designed to empower Gulf residents to take charge of their own recovery. Through Congressional action to dedicate any obtained CWA civil penalties to the Gulf and creation of a Gulf Coast Recovery Fund and Gulf Coast Recovery Council, the Gulf can begin to repair the damage and address the broader challenges facing the Gulf Coast's ecosystems and build upon existing mechanisms to address the negative economic and health effects caused by the spill. Although this funding may not provide for all the needs identified in this report, this report strongly recommends that Congress act quickly to dedicate a significant amount of any civil penalties recovered under the Clean Water Act from parties responsible for the spill to those directly impacted by the spill as a critical first step.

LONG-TERM RECOVERY WORKING GROUPS

The four chapters that follow were developed over the course of the past three months through the tireless efforts of four working groups. The work of these groups documents some of the impacts of the Deepwater Horizon oil spill on the people, economy, and ecosystems of the Gulf Coast and helped inform the final long-term recovery recommendations made in this report.

As previously mentioned, during the Deepwater Horizon Oil Spill the National Incident Command established several interagency teams in addition to the traditional National Contingency Plan response and clean up functions. The teams provided core staff to the four long term recovery working groups that have contributed to this report. The teams coordinated integration of private and governmental recovery assistance and have focused on the longer term issues such as health, social services, the economy, and environment and natural resources. These issue areas represent important federal responsibilities which are contingent upon the availability of funding. These key recovery support functions were identified by stakeholders during the development of the upcoming National Disaster Recovery Framework, which is designed to ensure early attention to recovery issues and a smooth transition from response to recovery.

Ecosystem Restoration Working Group

The long term ecosystem restoration working group was led by the Council on Environmental Quality and was composed of the departments of Agriculture, Commerce, Homeland Security, Interior, Justice, and Transportation, along with the Environmental Protection Agency, the Army Corps of Engineers, the Office of Management and Budget, and the Office of Science and Technology Policy. The working group drew upon previous work by these agencies from the Roadmap for Restoring Ecosystem Resiliency and Sustainability as well as discussions with state officials, stakeholder groups and state and local leaders in the environmental community.

Health Recovery Working Group

The health recovery working group was led by the Domestic Policy Council and the Department of Health and Human Services and included the Departments of Labor, Commerce, and Agriculture, along with the Environmental Protection Agency, the Council on Environmental Quality, the Office of Management and Budget, and the Office of Science and Technology Policy. The working group drew upon the work these agencies have been conducting on the Gulf Coast to ensure worker safety, seafood safety, long-term health monitoring, and behavioral health support.

Economic Recovery Working Group

The economic recovery working group was led by the National Economic Council and composed of the departments of Commerce (Economic Development Administration), Agriculture (Rural Development), Housing and Urban Development, Energy, Treasury, Labor, and Homeland Security, along with the Small Business Administration, the Office of Management and Budget, and the Council of Economic Advisors. The group drew upon previous work from the National Disaster Recovery Framework, as well as from the experiences of 21 economic-assessment teams that interacted with local communities across the Gulf Coast.

Corporation for National Community Service

The Corporation for National Community Service (CNCS) worked with the other three working groups to assess the contribution and needs of nonprofits working on the Gulf Coast, and to determine how nonprofit organizations can contribute to long-term recovery efforts.

LONG-TERM ECOSYSTEM RESTORATION

I. Introduction

The Deepwater Horizon oil spill underscores the critical linkage between the environment and the economic health of the Gulf of Mexico. There is a broad dependence of Gulf Coast communities on the sustained beauty and bounty of the region for their lives and livelihoods. This linkage was clearly demonstrated every day that oil continued to flow from the Deepwater Horizon wellhead. A critical component of the recovery from the spill is restoring the marine and coastal ecosystems of the Gulf, both for the public, and for those who rely on its natural resources for their livelihood.

The need to recover from the direct impacts of the oil spill should serve as the impetus to jumpstart a broader effort to restore and protect the Gulf Coast's ecosystems. This is a region that was already struggling with urgent environmental challenges, such as the loss of coastal barrier islands and marshes that protect coastal communities from frequent hurricanes. While trustee agencies are working to assess and repair the injuries to the environment from the Deepwater Horizon oil spill, it only makes sense to look at the broader challenges facing the system, and to leverage ongoing efforts to find solutions to some of the complex problems that face the Gulf. Sustained activities that restore the critical ecosystem functions of the Gulf will be needed to support and sustain the region's economic revitalization.

There are numerous ways to work toward achieving ecosystem restoration, including maintaining and restoring critical habitats; acquiring habitats for protection; taking regulatory action; and reducing flows of excess nutrients into the Gulf to improve water quality and fishery stocks. Ecosystem restoration will support economic vitality and employment; enhance human health and safety; protect infrastructure and communities; enable communities to better withstand storms; facilitate commerce and trade; sustain safe seafood and clean water; provide recreational and cultural opportunities; and contribute to the overall resilience of coastal communities and the nation.

The Gulf of Mexico encompasses dynamic, interconnected ecosystems spanning 600,000 square miles, with thousands of miles of shoreline, bayous, and bays across five U.S. states (Alabama, Florida, Louisiana, Mississippi, and Texas) and six Mexican states. Within the United States, the Gulf of Mexico watershed extends over a thousand miles upstream and drains over 40 percent of the contiguous U.S.[13] The region is highly diverse – both environmentally and culturally – and this diversity has made the Gulf region an economic engine for the nation. The Gulf of Mexico system contributes a range of ecosystem services – including fisheries, energy production, infrastructure protection, and recreational opportunities – that provide jobs and food, enhance safety, underpin the unique cultures of the region, and generate billions of dollars in revenue annually.

A number of environmental problems – the causes of which are both natural and manmade – have plagued the Gulf of Mexico. Though some of the problems are relatively state-specific, many of these problems span the entire region, and their solution requires a federal coordinating role to ensure that Gulf-wide activities are facilitated by coherent policies that promote science-based decision-making, balance objectives, and ensure the consistent, effective application of federal resources. The federal coordinating role is to promote a shared vision among the states and other stakeholders in planning and prioritizing actions that will sustain the many valuable goods and services that the Gulf of Mexico provides to coastal communities and the nation.

Ecological restoration is an intentional activity that initiates or accelerates the recovery of an ecosystem with respect to its health, integrity and sustainability. It also includes protecting and conserving healthy ecosystems so they can continue to provide services to reduce impacts from tropical storms and other disasters; support robust economies; and assist in mitigating and adapting to the impacts of climate change. Understanding restoration to mean restoration, enhancement, and protection is important for a comprehensive, proactive effort to enhance the resiliency of Gulf ecosystems and the ecosystem services they provide.

Recovery from the oil spill and the broader restoration of the Gulf region will take sustained leadership and commitment over many years. This is a partnership and a collaboration that requires participation by and investment from all parties – federal, state, tribal, and local governments; non-profit organizations; the private sector; academics; and scientists. It is also important to note that there are other activities in the coastal zone that could impact ecological goals and objectives, such as levee construction; dredging operations; and oil and gas activities. These activities must be carefully assessed and balanced with restoration goals.

Although the long-term goal should be to restore ecological processes, it is not possible to restore the ecosystem to a pristine pre-industrial state, in the face of climate change and other uses of the coast. The goal of the broader recovery is to reverse long-term degradation of the Gulf ecosystem's health.

It is important to recognize that some of the most significant problems affecting the Gulf cannot be solved within the five-state region alone. Coordination between ecosystem restoration efforts in the Gulf and work on stressors both upstream and outside of the U.S. is important. However, for the purpose of this chapter and the recommendations included here, the focus will remain on the five states bordering the Gulf of Mexico and the adjacent waters of the Gulf.

II. Background Information

The Gulf of Mexico region comprises the coastal areas of five southern states (Alabama, Florida, Louisiana, Mississippi, and Texas) and the offshore waters of the Gulf of Mexico extending into the exclusive economic zone of the United States.[14] The U.S. portion of the

Gulf of Mexico is fed by 20 river systems covering over 3.8 million square kilometers of the contiguous United States. The Mississippi River Basin alone drains over 40 percent of the contiguous U.S. and brings excess nutrient runoff from 31 states and two Canadian provinces into the Gulf.

The Gulf of Mexico is home to a diverse and vibrant network of interconnected ecosystems that, despite a variety of environmental stressors, offer immense productivity. The abundance and variety of life of the Gulf of Mexico and its coastal areas are an ecological, economic and cultural treasure. The Gulf's dynamic shoreline includes beaches, marshes, forested wetlands, and mangroves that provide aesthetic value; support a host of habitats and recreational uses; and provide the backdrop for the unique cultures and heritage of this region.

Some of the world's most important migratory bird flyways cross the Gulf of Mexico and a wide variety of bird species utilize the marshes, dunes, rivers, and other coastal habitats for foraging and nesting. The coastal wildlife refuges, parks, preserves, and other areas managed by public and private entities offer nurseries for many species of fish, invertebrates, and birds as well as opportunities for recreation. The Gulf region is home to a number of coastal, marine, and freshwater species listed as threatened or endangered under the Endangered Species Act, including five species of sea turtles. Twenty-one species of marine mammals in the Gulf (including dolphins, manatees and whales) are protected under the Marine Mammal Protection Act.

The coastal shelf and deeper habitats, as well as the water column above them, are home to rich biological assemblages of plants, animals, and microbes – ranging from deepwater coral, sponge, and fish communities on rocky reefs, to unique methane seep communities, and from the poorly characterized deeper parts of the water column to the rich planktonic assemblages in the upper waters exposed to sunlight. A plethora of commercial and recreational fisheries depend upon the health of this integrated ecosystem, including the coastal habitats that serve as nursery grounds for the vast majority of economically important species, including fish, shrimp, and crabs.

The Gulf Coast ecosystems are highly complex and dynamic and provide immense aesthetic, economic, and environmental value. Barrier islands and wetland complexes of this region provide critical infrastructure and are the first line of defense for coastal communities against powerful and persistent storms. They are also rich havens of biodiversity that provide essential habitat to a number of commercially and recreationally important species of fish, invertebrates, mammals, and birds. Moreover, the Gulf's wetlands provide a flood attenuation function, which reduces the impacts of flooding associated with storms, and they reduce potential future impacts associated with climate change. These ecosystems also perform critical water filtration functions, removing and trapping contaminants in the water and storing carbon. Collectively, the functions and services provided by these systems are considered key "ecosystem services" of the Gulf Coast.

A. PreExisting Ecosystem Challenges in the Region

The Gulf Coast has been and continues to be subject to a number of ongoing environmental challenges that have attracted significant attention from state and federal natural resource managers and conservation interests. Despite these problems, which are significant cause for concern, the Gulf continues to be intensely productive and supports multiple human uses. Ongoing issues include:

- *Loss of wetland habitats, including coastal marshes, forested wetlands, barrier islands, and coastal shorelines that form the Mississippi River Delta and Chenier Plains.* While an issue in every Gulf state, the loss of coastal habitat has been most dramatically illustrated in Louisiana, and highlights the need to maintain freshwater and sediment flows to the Gulf of Mexico. Since the 1930s, the coast of Louisiana has lost over 2,000 square miles (25-35 square miles per year) of wetlands.[15] Causes of this loss include a combination of erosion; storm damage; land subsidence; alterations of the natural freshwater and sediment flow from the Mississippi River; dredging of canals for oil and gas exploration and pipeline- installation activity; and the construction of navigation and flood control structures along the Mississippi River. Climate change (including the impacts of inundation and sea level rise) threatens to accelerate the loss of these habitats.
- *Erosion of barrier islands and shorelines throughout the Gulf Coast.* From Florida to Texas, continued erosion of the coastal barrier island system undermines storm protection for coastal communities, threatens the beaches that support the local tourism economy, and affects numerous species that rely on these barrier islands for habitat (e.g., Kemp's Ridley and loggerhead sea turtles, numerous shorebirds, and the Alabama beach mouse).
- *Loss and degradation of coastal estuarine habitat.* The estuaries and coastal systems of the Gulf Coast – such as Mobile Bay, Apalachicola Bay, Galveston Bay, Tampa Bay, Florida Bay, the Mississippi Sound, Barataria Bay, and others – provide the nursery habitat for most of the fishery resources in the Gulf, and support a nationally important oyster industry. These estuaries are impacted by a variety of stressors, including pollution, coastal development, energy development, erosion, hydrological alteration, changes in freshwater inflow, structural marsh management, and overfishing. Many of these bay systems have been recognized as estuaries of national significance by the National Estuary Program of the Environmental Protection Agency (EPA) under the Clean Water Act or as National Estuarine Research Reserves under the Coastal Zone Management Act, and are the focus of existing place-based protection and restoration efforts.
- *Imperiled Fisheries.* Several of the major commercially and recreationally important finfish species are currently experiencing pressures from overfishing or have been overfished. In some cases, these conditions have persisted for many years. Additionally, contaminants such as methyl-mercury in fishes, and red tide organisms and human pathogens in shellfish, reduce fishery values and endanger human health. Under the Magnuson-Stevens Fishery Conservation and Management Act, federally managed populations of National Oceanic and Atmospheric Administration (NOAA) trust species such as red snapper, grouper, and mackerel are being rebuilt through the efforts of the Gulf of Mexico Fishery Management Council and state fishery management agencies.[16] The impact of the Deepwater Horizon oil spill on the rebuilding efforts for these species is as yet unclear.
- *Hypoxia (low oxygen) in the Gulf of Mexico.* Hypoxia occurs where the concentration of dissolved oxygen in the water column decreases to a level that reduces the quality of habitat, resulting in death or migration away from the hypoxic

zone. The northern Gulf of Mexico adjacent to the Mississippi River is the site of the largest hypoxic zone in the United States and the second largest hypoxic zone worldwide.[17] This Gulf of Mexico "Dead Zone" is caused by input of excess nutrient pollution to the Gulf, most of which comes from upstream through Mississippi River drainage. A federal-state Hypoxia Task Force has been working to address factors leading to low-oxygen conditions, and EPA and the Department of Agriculture (USDA) have jointly worked to develop strategies to reduce nutrient runoff. NOAA has been working on developing models to support better understanding of biological-systems transport, including hypoxia predictions.

- *Climate change.* Our changing climate is already altering, perhaps irreversibly, the physical, chemical, and biological characteristics of our oceans, coasts, and adjacent watersheds. Increasing air and water temperatures, changing precipitation patterns, rising sea levels, and ocean acidification will increasingly confound efforts to restore or sustain system states. Federal and state natural resource managers need the information and tools to develop strategies for mitigating and adapting to a dynamic environment and ongoing habitat reorganization and restructuring.

The degradation of coastal and marine ecosystems and the services they provide has direct and indirect impacts on the economy, communities, and environment of the region. Future impacts associated with storms, subsidence, and sea level rise will serve to only amplify the region's vulnerability. Steps must be taken to curb the impacts of multiple stressors on Gulf ecosystems in order to reduce the negative consequences for the marine and terrestrial environment, national commerce, maritime industry, energy security, fisheries, and the rich cultural legacy of the Gulf Coast.

B. How the Deepwater Horizon Oil Spill Further Impacts the Region

The Deepwater Horizon oil spill has caused significant environmental impacts. Federal and state agency scientists and independent researchers are currently working to understand and quantify the injuries to natural resources, ecosystems, and their services caused by the Deepwater Horizon oil spill. From an ecological standpoint, four areas of great concern are the impacts on the water column, fisheries, habitats, and species.

Water Column Impacts

It is estimated that the Deepwater Horizon oil spill has released over 4.9 million barrels of oil into the Gulf of Mexico.[18] This event is unique in its magnitude in volume and area but also in its location. The spill released the oil, and the response included the addition of dispersants, at almost a mile depth. This part of the open ocean is subject to strong and variable ocean currents that can transport oil, dispersants, and tar balls to remote locations, sensitive and valuable ecosystem areas, areas of large commercial and recreational use, and close to highly populated coastal areas. The eventual degradation of the oil and its constituents will occur at fine scales (through microbial processes) and may have significant impacts on the food chain. Monitoring of physical, chemical, and biological factors in the whole water column, at the oil spill site, and away from the wellhead is necessary to fully comprehend the impact of the oil spill on ecosystems and coastal areas.

Fisheries Impacts

The Deepwater Horizon oil spill has had a negative environmental, economic, and social impact on Gulf fisheries due to oil-related closures and concerns about seafood contamination. The extent of the harm is not yet well understood. Resource managers and scientists continue to discuss possible short- and long-term impacts to key stocks, indicator species, and food webs, but the studies required to address these issues are still in the earliest stages.

The Deepwater Horizon oil spill forced the temporary closure of up to 88,522 square miles or 36 percent of federal Gulf waters[19], and more in-state waters, to fishing. The majority of state and federal waters have been reopened based on joint efforts of NOAA, the EPA, the Food and Drug Administration (FDA) and the states in developing a reopening protocol that includes sensory and chemical testing of seafood for components of the oil. Finfish, shrimp, crabs, and oysters have been tested and were found to be safe for human consumption.

Fish eggs and larvae are particularly sensitive to even very low levels of oil in water. Crabs and oysters, which lack the ability or tendency to swim away from waters that are contaminated with oil, are also vulnerable.

The impact of the oil spill also extends to fresh-water ecosystems. Diadromous fishes move between fresh water and salt water to complete their life cycle and, therefore, are influenced by conditions in both environments.

Habitat Impacts

The Gulf of Mexico ecosystem hosts a variety of significant coastal and marine habitats, some of which are in marine protected areas that have been exposed to oil. These vital habitats include beaches; wetlands (including marshes, forested wetlands, mangroves, seagrasses and other submerged aquatic vegetation); intertidal and subtidal mudflats; oyster-reef habitats; shallow and deepwater corals; estuarine and marine benthic habitat; and marine habitat.

The long-term effects of oil on these habitats have yet to be determined. Oil can persist in wetlands and soft sediments for years and kill plants and animals through smothering or chemical toxicity. If there are significant impacts on marsh grasses, mangroves, and other vegetation, this could accelerate erosion of marsh edge, which would only exacerbate a large-scale problem plaguing the region. In addition, oil that becomes entrained with suspended sediment or other particles can sink to the bottom, affecting submerged vegetation, oysters, coral, and other benthic species that live on the seafloor.

Species Impacts

The nature of the impacts and their geographic and temporal scope are expected to vary by species. Many animals have died from direct contact with the oil, but more insidious, long-term stress from the oil and clean-up activities may affect food availability, growth, reproduction, behavior, disease, and the long-term health of many organisms.

Sea turtles, all of which are either endangered or threatened, feed along floating debris and seaweed (including *Sargassum*) that were heavily oiled during the spill. *Sargassum* is considered essential fish habitat for the juvenile life stages of a number of highly important recreational and commercial species, including mahi mahi, billfish (marlin and sailfish), and cobia. Additionally, the coastal lands and waters off Louisiana, Mississippi, and Alabama include critical habitat for species designated under the Endangered Species Act.

The Gulf of Mexico is the only spawning ground of the critically depleted western Atlantic population of the bluefin tuna, which was just past the peak of its spawning season at the time of the spill. Tuna produce buoyant eggs, which may be particularly impacted by the oil spill. Other threatened species, such as Gulf sturgeon and several species of shark, may also be impacted by the spill. Other impacts may be identified in the coming years.

C. Initial and Current Response to the Oil Spill

The Deepwater Horizon oil spill added new stresses on the Gulf of Mexico region's ecosystem and economy. The spill's impacts include both obvious, dramatic short-term impacts – such as the closure of up to 36 percent of federal waters in the Gulf to fishing during the oil spill, the oiling of birds, other animals, beaches, and coastal marshes, the disruptions associated with the massive response efforts in the Gulf and on-shore – as well as long-term impacts.

A key statute guiding the government's response to this latest disaster is the Oil Pollution Act of 1990 (OPA) and its related regulations. Passed in the wake of the 1989 *Exxon Valdez* oil spill, OPA expanded the scope of the National Oil and Hazardous Substances Pollution Contingency Plan, more commonly called the National Contingency Plan (NCP), by mandating new contingency planning and response preparedness responsibilities for the federal government and industry, as well as by providing additional guidance on coordinating and directing response and cleanup activities.

In addition, OPA provided funding for oil spill related response activities through the Oil Spill Liability Trust Fund, amended the Clean Water Act to increase penalties for oil discharged, broadened the response and enforcement authorities of the federal government; and preserved state authority to establish laws governing oil spill prevention and response. OPA also set financial responsibility requirements, and subjected parties determined responsible for the release or threat of release of oil to liability for a variety of removal costs and damage claims.

Under OPA, the process for restoring the damage to natural resources has already begun. A major goal of the legislation is to compensate the public for injury to, destruction of, loss of, or loss of use of, natural resources, including the reasonable costs of assessing the harm done as a result of a discharge or substantial threat of a discharge of oil.

This goal is achieved through returning injured natural resources and services to the condition they would have been in if an event had not occurred (otherwise referred to as "baseline" conditions), and compensating for interim loss of use of those resources, which are measured from the date of the injury until such natural resources are restored.

Under OPA Natural Resource Damage provisions, responsibility for assessing injuries to natural resource as a result of an oil spill, and implementing a restoration plan intended to compensate the public for those injuries, is shared by federal and state "trustee" agencies designated by the president and the governor of each affected state, and by Indian tribal trustees when natural resources under a tribe's jurisdiction have been injured. There are usually multiple trustees with overlapping responsibility for injured natural resources, and they typically seek to coordinate their efforts through a Trustee Council.

OPA includes a process referred to as Natural Resource Damage Assessment, or NRDA, which the trustees use to:

- Identify injuries to natural resources and services resulting from an incident.
- Provide for the return of injured natural resources and services to baseline conditions.
- Compensate for the losses that continue to occur until baseline conditions are returned.
- Include public involvement in the restoration planning process.

The NRDA process is governed by federal regulations, which define three main phases of a N R DA:

- Pre-assessment Phase, in which the trustees determine if they have the jurisdiction to pursue restoration under OPA and, if so, whether it is appropriate to do so.
- Restoration Planning Phase, in which the trustees evaluate the potential injuries to natural resources and services and use that information to determine the need for and scale of restoration actions through two basic components: injury assessment and restoration planning.
- Restoration Implementation Phase, in which the Final Restoration Plan is presented to parties responsible for the spill to implement or fund the trustees' costs of implementing the plan. If the responsible parties decline to pay the claim, OPA authorizes trustees to bring a civil action for damages in federal court or to seek payment from the Oil Spill Liability Trust Fund for such damages (and allow the Fund to recover payment from the responsible parties as appropriate).

Funds recovered from responsible parties under OPA must be used for restoration, rehabilitation, replacement, or acquisition of equivalent natural resources or services. During the Deepwater Horizon oil spill, federal and state entities with natural resource trust responsibilities initiated the NRDA process to assess natural resource injuries caused by the spill, and to identify appropriate restoration options.

The natural resource trustees have formed a Trustee Council that presently includes representatives from Alabama, Florida, Louisiana, Mississippi, and Texas, and for the federal government the leads are the Departments of Commerce and Interior. As part of the NRDA process, the Trustees are reviewing and, as appropriate, incorporating the vast amount of monitoring data on the Gulf of Mexico into their assessment of the injuries that may potentially result from the Deepwater Horizon oil spill.

The Trustees also intend to develop a restoration plan; they will review and develop a suite of potential restoration options that ultimately will be scaled directly to the quantified injuries and will form the foundation of the natural resource damage claim. During the Trustee Council's process of developing the restoration plan, the Oil Pollution Act requires public review and comment.

The long-term Gulf of Mexico ecosystem restoration effort described throughout this report must be closely coordinated with the NRDA Trustee Council, which is responsible for management of the NRDA for the Deepwater Horizon oil spill. The two efforts will share information, and the long-term ecosystem restoration effort will seek to build off and complement the progress made through the NRDA process under OPA, but will not interfere with the Trustee Council's exercise of its statutory responsibilities. For example, the mechanisms proposed in Section IV could identify projects that might meet the needs of the eventual NRDA restoration plan.

In addition to response efforts under OPA, federal agencies and private partners took early action during the oil spill to avoid and minimize impacts to wildlife. A partnership between the National Fish and Wildlife Foundation, USFWS, NOAA, and private partners took action to transfer of 25,000 sea turtle eggs from Gulf of Mexico beaches to the Atlantic coast of Florida. The Natural Resource Conservation Service worked with landowners in the Gulf Coast to enhance migratory bird habitat to keep birds away from oiled areas. Conservation organizations throughout the region engaged in complementary efforts with the support of the National Fish and Wildlife Foundation.

D. Major Relevant Efforts in the Region

At the time of the spill, there were already numerous entities engaged in restoration and recovery efforts in the Gulf region, including federal agencies; state and local governments; tribes; private corporations; non-governmental organizations (NGOs); academic and scientific organizations; private landowners; affected stakeholders; and the public. In addition, this administration has undertaken a number of initiatives across the region to coordinate federal programs and policies with local efforts. Many of these efforts seek to construct restoration projects on-the-ground and in-the-water and to operate under an array of authorities, programs, and management objectives.

The following summaries provide an overview of some of the major efforts underway in the Gulf of Mexico and on the Gulf Coast. These efforts represent critical first steps in restoring Gulf ecosystems, and this report seeks to build upon and coordinate with these efforts and draw from their expertise.

A review of existing programs indicates that there should be a mechanism to ensure communication and integration, in order to avoid duplication of efforts. However, this report is not intended to impede or derail any of these other programs. Rather, where possible, these existing efforts should be accelerated in an effort to achieve results. For example, existing efforts on hypoxia, including work by EPA and USDA to reduce nutrient loadings to the Mississippi Basin, can and should be accelerated during this effort.

State and Federal Restoration Efforts

The Gulf states, federal government, and localities have all engaged in numerous state-specific and local restoration planning efforts for the last several decades. These efforts vary in their geographic and ecological scope. These state-specific and local efforts will help drive priorities and identify critical gaps in the broader Gulf-wide restoration effort. The following non- exhaustive list describes several of the more prominent programs by state:

- Alabama: The state of Alabama has focused on a variety of restoration and conservation activities in its coastal areas. It cites tax revenues generated through tourism activity in the coastal counties as an important contribution to the state economy. Barrier-island restoration is a priority for the state. The health of Mobile Bay, particularly the Mobile- Tensaw River Delta, has been a major focus of state efforts. The Mobile River basin drains an area of nearly 44,000 square miles across Alabama, Mississippi, Tennessee, and Georgia into Mobile Bay.[20] Through the Forever Wild Program, and other programs, the state has invested in land protection around the delta and is working to address hydrologic connections, water quality, and nutrient pollution in the bay[21].

- Florida: A variety of state, local, and federal partners have been working together for decades on a number of significant scientific, planning, and implementation efforts to protect and restore the Everglades, including the implementation of the Comprehensive Everglades Restoration Plan (CERP). Other statewide programs include the Coastal Wildlife Conservation Initiative, a multi-agency effort led by the Florida Wildlife Commission to address threats to native wildlife and their habitats in coastal ecosystems. The Statewide Beaches Habitat Conservation Plan, led by the Department of Environmental Protection, seeks to preserve the unique and precious wildlife and natural resources of Florida's coastline.
- Louisiana: Due to a high rate of land loss and the impacts of Hurricanes Katrina and Rita in 2005, Louisiana has engaged in robust planning efforts that include the State Comprehensive Master Plan for a Sustainable Coast (Master Plan). This effort was completed in 2007, and outlines a dynamic, multipurpose strategy to integrate planning, design, and implementation of hurricane protection projects with projects aimed at restoring Louisiana's rapidly eroding coastal wetlands.

  The Master Plan serves to guide all coastal restoration and hurricane protection efforts in Louisiana over the next several decades. The state is currently updating its Master Plan for 2012. The federal government has also made coastal restoration in Louisiana a priority, with efforts that include the Louisiana Coastal Area (LCA) Near-Term Plan and the Louisiana Coastal Protection and Restoration Plan (LACPR). Finally, the federal Coastal Wetlands Planning, Protection, and Restoration Act (CWPPRA) represents one of the oldest and largest federally funded restoration programs in the country.
- Mississippi: Mississippi has engaged in an active federal-state restoration effort called the Mississippi Coastal Improvement Program (MsCIP), which was developed to analyze and design comprehensive improvements or modifications to existing improvements in the coastal areas of Mississippi to reduce storm damage, prevent saltwater intrusion, conserve fish and wildlife, prevent erosion, and for other purposes. Fifteen "interim" projects that were congressionally authorized and funded in 2006 following Hurricane Katrina are included in MsCIP, with most of those fifteen projects now either underway or complete. In 2009, Congress authorized and appropriated $439 million for the U.S. Army Corps of Engineers to conduct barrier island restoration and other restoration opportunities. Since the final MsCIP report was recently completed and sent to Congress, the remaining projects have not yet been authorized.
- Texas: In Texas, coastal restoration priorities have focused on a number of areas including estuaries, coastal barrier islands, sea turtle conservation, and coastal bird habitats. Similar challenges to those found elsewhere in the Gulf - sea level rise, erosion, subsidence, and the need to restore hydrologic connections - are priorities for the state. For example, Texas has identified concerns with declines in coastal waterbirds in the mid-coast area and is working to restore eroding or subsiding habitat. In Galveston Bay, Hurricane Ike impacted nearly 8,000 acres of oyster reef and the state is working with partners to remove silt and restore these areas.[22] Finally, the state is working to protect and restore nesting beaches for Kemp's Ridley sea

turtles to provide critical backup nesting sites. Recognizing declines in seagrasses within Galveston Bay and some Coastal Bend regions of Texas, the state developed a comprehensive Seagrass Conservation Plan for Texas in 1996 to guide concerted planning and implementation of conservation and restoration actions to address documented loss and degradation throughout the state's bays and estuaries. Texas also has a state-wide Coastal Erosion Protection Planning and Response Act Program.

Gulf of Mexico Alliance (GO MA)

The Gulf of Mexico Alliance (GOMA) is a partnership between Alabama, Florida, Mississippi, Louisiana and Texas that was founded in 2004. GOMA engages federal agencies, academia, and non-governmental partners to develop strategies and coordinate implementation of the Governors' Action Plan. The Action Plan identifies regionally significant issues that could be effectively addressed through increased collaboration at the local, state, and federal levels by building partnerships and laying a foundation for a regional approach to ecosystem management and restoration. The Action Plan identifies six priority areas: 1) water quality for healthy beaches and seafood; 2) habitat conservation and restoration; 3) ecosystem integration and assessment; 4) reducing nutrients' impacts on coastal ecosystems; 5) coastal community resilience; and 6) environmental education.

Hypoxia Task Force

The Mississippi River/Gulf of Mexico Watershed Nutrient Task Force, more commonly called the Gulf Hypoxia Task Force, provides executive level direction and support for coordinating the actions of ten states and five federal agencies working on nutrient management within the Mississippi River/Gulf of Mexico Watershed. Co-chaired by the EPA and the state of Mississippi, the task force oversees the implementation of the 2008 *Action Plan*. The *Action Plan* outlines critical needs to complete and implement nutrient reduction strategies, promote effective conservation and management practices, track progress, reduce existing scientific uncertainties, and promote effective communication to increase awareness of Gulf hypoxia.

Roadmap for Restoring Ecosystem Resiliency and Sustainability

In October 2009, President Obama formed the Louisiana-Mississippi Gulf Coast Ecosystem Restoration Working Group, co-led by the Council on Environmental Quality (CEQ) and the Office of Management and Budget (OMB) and comprising high-level political appointees from the National Oceanic and Atmospheric Administration, the U.S. Environmental Protection Agency, and the Departments of the Army, Homeland Security, the Interior, and Transportation. In March 2010, the Working Group released a *Roadmap for Restoring Ecosystem Resiliency and Sustainability* (Roadmap). The Roadmap reaffirmed the federal government's partnership and coordination with the states of Mississippi and Louisiana. It outlined federal actions over an 18-month period to address pressing near-term policy, process, and legal hurdles to restoration and laid the foundation for a long-term comprehensive vision to be achieved jointly with the states.

The Roadmap was developed to guide near-term federal actions for the restoration and protection activities in Louisiana and Mississippi, and seeks to complement and support the longer-term ecosystem restoration planning activities already under way in those states and in the various federal agencies. Recognizing that providing for the sustainability of the region is something that will take a committed effort over many years, the Roadmap outlines a process to establish a new and enduring entity to provide long-term leadership, shared responsibility, and coordination to resolve the difficult challenges facing the region. Establishment of a Gulf Coast Ecosystem Restoration Task Force described in this chapter could fulfill the need for a long-term coordinating body described in the Roadmap, while other actions outlined in the Roadmap could go to inform the work of the task force.

National Ocean Council and Regional Planning Bodies

In July 2010, President Obama signed Executive Order (EO) 13547, which creates a new National Ocean Council to coordinate federal stewardship efforts in the ocean, our coasts, and the Great Lakes. The National Ocean Council has nine National Priority Objectives, including "Regional Ecosystem Protection and Restoration," and as this new body identifies particular regions on which to focus, it is possible that the Gulf could become one of those regions. As that effort develops over the coming year, we should evaluate and consider the connections between that effort and the long-term restoration work outlined in this report.

EO 13547 also outlines a new approach to managing the multiple uses of coastal waters, referred to as "coastal and marine spatial planning." Coastal and marine spatial planning moves away from the current sector-by-sector, statute-by-statute approach. It moves toward management that can properly account for cumulative effects, sustain multiple ecosystem services, and explicitly evaluate the tradeoffs associated with proposed alternative uses.

Under this new approach, the National Ocean Council will work with states and federally recognized tribes to create regional planning bodies that guide the development of regionally driven plans. One of the National Ocean Council's designated regional planning areas is the Gulf of Mexico, which covers all five states discussed in this report. The membership of each regional planning body, which will be established by early 2011, will consist of federal, state, and tribal authorities relevant to coastal and marine spatial planning for that region.

Climate Science and Landscape Conservation Cooperatives

Several Gulf states have been working in cooperation with federal agencies, including the Department of the Interior, as part of a national initiative to develop Climate Science Centers (CSCs) and Landscape Conservation Cooperatives (LCCs). The CSCs will provide fundamental scientific information, tools, and techniques that land, water, wildlife, and cultural resource managers and other interested parties can apply to anticipate, monitor, and adapt to climate change impacts. As they are developed in the Gulf, LCCs will focus principally on coastal and near-shore areas and will be applied conservation science partnerships that are intended to drive success at landscape scales. LCCs have a role in helping partners identify common goals and priorities to target the right science in the right places for efficient and effective conservation. LCCs provide scientific and technical support for landscape-scale conservation in an adaptive management framework by supporting conservation planning, decision support tools, prioritizing and coordinating research, and designing inventory and monitoring programs.

In addition, NOAA's climate services directors will build and strengthen regional partnerships to better assess and deliver regionally-focused climate science and information products and services to help people make informed decisions in their lives, businesses and communities.

Coastal and Marine Managed Areas

There are several federal programs in the Gulf region that have an important role addressing coastal and marine issues, including habitat restoration and protection, pollution, hypoxia, and climate change, through long-term research and monitoring, outreach and education, and environmental stewardship. The National Estuary Program (NEP), established by Congress in 1987 as part of the Clean Water Act, works to restore and maintain water quality and ecological integrity of estuaries of national significance. Seven of the 18 NEPs are located in the Gulf region. The National Estuarine Research Reserve System (NERRS), established by the Coastal Zone Management Act in 1972, is a series of 27 coastal areas in the United States, with five in the Gulf of Mexico. These reserves are protected for long-term research, water-quality monitoring, education, and coastal stewardship. Under the mandates and authorities of the National Marine Sanctuaries Act of 1972, NOAA manages 13 national marine sanctuaries, of which two (the Florida Keys and Flower Garden Banks) are in the Gulf of Mexico. The protection of these nationally significant marine areas is important for maintaining, protecting and restoring habitats and ecosystem services.

E. The Ecological Economic Health Nexus

The resources of the Gulf of Mexico serve as an important economic driver for both the region and the United States, as well as the cultural identity for many communities in the region. As discussed in detail in the economic chapter's section on background information, commercial, recreational, and subsistence fishing, oil and gas exploration and extraction, shipping, recreation, and tourism traditionally are the main drivers of the Gulf economy. The economy of the region is largely dependent on the health of the ecosystem and the safety of seafood, meaning that environmental impacts, such as an oil spill, can have serious economic impacts on millions of people. Likewise, efforts to improve the ecosystem in the Gulf region have important feedback effects on the region's economic health.

Beyond the core Gulf of Mexico jobs in tourism, fisheries, and other natural resource dependent sectors, ecosystem restoration offers "green jobs" opportunities, including engineers, surveyors, and demolition and construction jobs. Coastal restoration projects create approximately 20 jobs for every $1 million spent. The average amount spent on salaries in proportion to the total project cost varies from project to project.

Any lost income in the region may result in diminished resources for volunteer organizations and charities that are augmenting state and local government efforts to support the provision of health-related prevention activities and services. This is particularly challenging given the behavioral health and human services needs that have increased as a result of the spill. From the fishing community to the local tourism industry, individuals in affected areas face an uncertain economic future, causing significant anger and resentment. The Deepwater Horizon oil spill has disrupted the delicate social, economic, and psychological balances in communities across the region.

Beyond resource extraction and management, Gulf of Mexico ecosystems provide a variety of ecosystem services that contribute to and protect the economy and culture of the region. These ecosystem services provide food and fiber, clean air and water, storm protection, recreation, carbon sequestration, and aesthetic values. Additionally, the coastal landscape of this region has sustained unique cultures for hundreds of years and ecosystem restoration and protection is critical to safeguarding and sustaining the region's rich cultural histories and heritages. Scientific research and modeling show that coastal wetlands, barrier islands, mangroves, and forested wetlands serve as natural buffer zones, protecting communities and other infrastructure against storms and floods and providing millions of dollars in savings annually.

III. Principles for Ecosystem Restoration

To restore the Gulf region, the restoration and recovery effort must recognize that the problems afflicting the Gulf of Mexico are of a great scale and the region has been significantly affected. Restoration and recovery efforts should recognize that that the economic vitality of the region is directly tied to the health of Gulf ecosystems. Under that premise we propose the following vision for the Gulf of Mexico ecosystem restoration:

Resilient, healthy Gulf of Mexico ecosystems that support the diverse economies, communities, and cultures of the region.

To truly create a resilient Gulf of Mexico, the federal government must assist state and local governments, nonprofit organizations, community groups, scientists, academics, the private sector, and the public to identify and address the ecological decline that has already negatively impacted the region, including significant habitat loss (wetlands, barrier islands, bottomland hardwood forests, etc.), decline of indicator species (species whose relative well-being in a given environment is indicative of the health of its ecosystem as a whole), and hypoxia. This process must be a truly collaborative partnership, and it will require participation by and investment from all parties listed above.

Restoring the natural processes of Gulf ecosystems so that they are healthier, and more resilient may take decades. Restoration should be science-based, and must include restoration of the natural processes that maintain healthy ecosystems. Important progress can and must be made now. This is an opportunity to address the root causes of Gulf degradation and to come up with creative solutions to address problems.

The following principles serve as the drivers for achieving the vision of resilient and healthy Gulf of Mexico ecosystems. These principles represent broad objectives that draw from, and build upon, existing plans for the Gulf described earlier in this document. For the restoration effort to succeed, it should be guided by science and include a clear set of measurable objectives tied to indicators that gauge progress toward achieving the restoration vision.

- *Principle 1: Coastal Wetland and Barrier Shoreline Habitats are Healthy and Resilient.* In order to sustain the many ecosystem services upon which humans rely, coastal habitats must be healthy and resilient. Reversing ongoing habitat degradation and preserving the remaining healthy habitats is a key principle. It must be recognized that even the healthiest ecosystems are dynamic, so a restoration effort should not focus entirely on a fixed "footprint." A key objective of this principle is to bring greater balance to managing the Mississippi River and other rivers for flood control, navigation, and ecosystem restoration. Another objective is to retain sediments that nourish coastal wetlands, rather than releasing them to the Gulf.

- *Principle 2: Fisheries are Healthy, Diverse and Sustainable.* The Gulf is home to the largest commercial fishery in the contiguous United States. The total trip expenditures for recreational fishing in the Gulf states in 2008 were nearly $1.5 billion.[23] Key objectives of this principle may include incorporating testing and other mechanisms for seafood safety to ensure that fish and shellfish are safe for human consumption, and working through regulatory and other conservation mechanisms to restore populations of fish and shellfish.
- *Principle 3: Coastal Communities are Adaptive and Resilient.* The needs and interests of Gulf communities vary and the most effective solutions will be based on local conditions. Given that much of the land affected by the oil spill is privately held, full restoration will rely on local citizen support. The impacts of climate change, including sea level rise and more frequent and intense storms, will likely alter the landscape significantly, forcing communities to reassess their priorities. Key objectives of this principle may include providing coastal managers with information and tools to make better land use and public health decisions, and increasing awareness of the connection between ecosystem and community resilience.
- *Principle 4: A More Sustainable Storm Buffer Exists.* Persistent coastal land loss, compounded by sea level rise, is deteriorating natural lines of defense, leaving coastal communities vulnerable to tropical storms. Natural and engineered systems are necessary to reduce exposure and ensure protection. Key objectives of this principle may include maintaining and expanding natural storm buffers such as wetland and barrier islands, improving decision-making with regard to structural protection and navigation interests so that these complement and enhance restoration of natural systems. Another objective is the reduction of risk posed to people and private property through effective planning, mitigation, and balancing of interests.
- *Principle 5: Inland Habitats, Watersheds and Off-Shore Waters are Healthy and Well- Managed.* Communities across the nation rely on our ability to maintain healthy, resilient, and sustainable ocean, coasts, and Great Lakes resources for the benefit of present and future generations. Additional stressors on the health of these systems and the resources they support include overfishing, pollution, and coastal development. Further, ocean and coastal resources are directly and indirectly impacted by land management and use decisions in the watersheds that drain into the Gulf of Mexico. Key elements of this principle include improving management of agricultural and forest lands; restoring floodplains and wetlands to improve water quality, reduce flood risks, and enhance wildlife habitat; reducing erosion and nutrient runoff from agricultural and developed land; and using state-of-the-art planning tools to deliver comprehensive, integrated ecosystem-based management of resources.

Each principle and its associated objectives will be based on science that will support the efforts designed to achieve them. To help achieve these objectives, agencies should work to develop performance indicators to evaluate effectiveness; adaptively manage efforts; provide a mechanism for oversight and accountability; and ultimately enable the vision of resilient and healthy Gulf of Mexico ecosystems.

IV. Recovery Strategy

From prior and ongoing planning and restoration efforts in the region, it has been learned that the best results are achieved through effective cooperation and coordination among various federal, state, and local entities. A task force where federal and state agencies can work together to coordinate their efforts, engage stakeholders, and address the policy and process obstacles to implementation will help align and advance restoration plans and projects. Below are the main elements of a Gulf Coast Ecosystem Restoration Task Force and Strategy that may be established under existing authorities that will provide the means to address some of the most pressing coordination needs in the region and that may evolve and adapt in the future.

Purpose and Need for a Gulf Coast Ecosystem Restoration Task Force and a Gulf Restoration Strategy

The scope and complexity of the sweeping ecological challenges facing the Gulf of Mexico region require action on a large scale. There is a need for an intergovernmental Gulf Coast Ecosystem Restoration Task Force that facilitates coordination of federal, state, and tribal efforts on long-term restoration and recovery. While the Ecosystem Restoration Task Force is not meant to direct other Federal agencies or the expenditure of funds, the purpose of the task force would be to serve in an advisory capacity to:

- Provide for the development of a Gulf of Mexico Regional Ecosystem Restoration Strategy.
- Help coordinate the selection, funding, and implementation of protection and restoration projects, in coordination with existing authorities and efforts, including the Deepwater Horizon oil spill NRDA Trustee Council.
- Ensure transparency and accountability in decision making.
- Enhance public engagement in development and implementation.
- Coordinate science (including monitoring and adaptive management), regulatory decision making, management of other types of projects (i.e., navigation and flood control) with ecosystem restoration to improve compatibility with a healthy and resilient coast.

With so many existing planning efforts, ongoing restoration projects and new initiatives that have emerged prior to and since the Deepwater Horizon oil spill, there is a critical need for a unifying coordination structure that can serve to coordinate, leverage, and accelerate these parallel paths.

A Gulf Coast Ecosystem Restoration Task Force should be established immediately under existing authorities to provide a forum to coordinate the implementation of existing authorities for ecosystem restoration in the Gulf of Mexico. Recommendations regarding membership, leadership, responsibilities, and near-term deliverables of the task force are described below. The membership, leadership, and responsibilities of this task force could change over time.

Membership and Leadership

The Gulf Coast Ecosystem Restoration Task Force would include federal, state, and tribal government representatives, engaged at a high level (Assistant Secretary or equivalent), with a director and a staff that is assigned from member agencies to support the work of the entire effort.

Federal membership on the task force would include the Departments of Agriculture, Commerce, Interior, Justice, and Transportation, along with the Environmental Protection Agency, U.S. Army Corps of Engineers, the Council on Environmental Quality, the Office of Science and Technology Policy, and the Domestic Policy Council. The Departments of Health and Human Services, Homeland Security, Labor, and the Small Business Administration would serve in an advisory role to offer guidance on issue-specific topics. All federal agency participation would be consistent with each agency's existing statutory and regulatory responsibilities.

Each Gulf state would have one representative on the task force, nominated by each Gulf state governor and appointed by the president. In many cases, individuals who would serve on this task force from the state agencies are likely to be the same as those serving as their state's representative on the Deepwater Horizon oil spill NRDA Trustee Council. Such overlap in membership would be encouraged as it could help ensure coordination with NRDA restoration activities. It is important that the task force reach out to federally-recognized tribes in the region to engage them in undertaking its responsibilities, in collaboration with them, and determine an appropriate structure for tribal participation.

The task force would be chaired by the Federal Gulf Coast Ecosystem Restoration Task Force chair, who would be appointed by the president and act as federal lead for Gulf recovery. There would also be one state vice-chair who would be jointly selected by the state representatives on the task force.

Responsibilities of the Task Force

The principal responsibility of the task force would be to coordinate intergovernmental efforts to implement restoration programs and projects in the Gulf of Mexico. This should include priority setting and better coordination of restoration efforts with federal and state agency regulatory and permitting authorities. The task force would develop ways to make restoration project implementation more coordinated and efficient, recognizing that many of the restoration needs in the Gulf of Mexico are of such scale and complexity that rapid implementation will not always be possible.

Still, efficiencies gained from improved coordination and communications among agencies are essential to smooth the path for critical projects. The task force would develop and foster innovative and collaborative ways to identify, design, review, implement, and assess restoration activities. The task force obviously cannot oversee implementation of every project, nor should it. Instead, this group should focus its project implementation efforts prioritizing and integrating efforts on complex, large-scale projects and on developing innovative approaches to resolving problems and conflicts. It would be a priority for this task force to identify policy and process barriers to implementation of restoration projects identified by task force members, as well as to address needs raised by the Deepwater Horizon oil spill NRDA Trustee Council.

Additionally, the task force should consider other mechanisms to communicate more broadly with interested constituents, building on and learning from the infrastructure that was set up to communicate information to the public during the response phase of the spill.

These mechanisms may include open, public meetings, a regularly updated website, conference calls with various stakeholder groups, and newsletters. The task force should also utilize existing meetings and events of stakeholder organizations, when possible, as opportunities to reach constituents.

The Gulf Coast Ecosystem Restoration Task Force also would be expected to coordinate with the Department of Health and Human Services on public health issues and with the departments of Commerce, Labor, and Treasury, along with the Small Business Administration and other federal agencies on ways to maximize the economic impact of ecosystem restoration projects.

Developing a Strategy for Gulf of Mexico Restoration

Within its first year, the task force would prepare a Gulf of Mexico Regional Ecosystem Restoration Strategy (hereafter, "the Strategy") that would identify major policy areas where coordinated federal-state action is necessary, such as sediment management. The Strategy would consider existing restoration planning efforts in the region to identify planning gaps and restoration needs, both on a state-by-state basis and on a broad regional scale. The members of the task force would identify ways under their various authorities to address these planning needs or propose new programs or actions needed to implement elements of the Strategy.

In accordance with the Strategy, the task force would work with the federal and state agency members on improving the alignment of each agency's relevant programs and authority with the Strategy. The task force would coordinate with OMB in the development of annual budget guidance on Gulf of Mexico priorities for all task force member agencies consistent with the Strategy.

The task force would work with existing federal and state advisory committees, as appropriate, to ensure that relevant scientific and technical knowledge underpins restoration planning and decision making, and that research, monitoring, and assessment efforts are organized to achieve the Strategy.

The task force would advise on Gulf of Mexico restoration efforts. Specifically, the task force would help agencies develop performance measures and prepare a biennial report card on progress toward the goals outlined in the Strategy. Implementation of Gulf-wide monitoring would provide the information necessary to evaluate progress toward performance measures. Furthermore, the task force could advise in the development of an overarching environmental review document under the National Environmental Policy Act in parallel with development of the Strategy from which individual project reviews would be tiered, as appropriate.

Outreach and Engagement

The task force would conduct its operations in such a way as to ensure that stakeholder groups, particularly those in the region, have access to share their needs and viewpoints, including state-specific perspectives. A wide variety of stakeholder groups should be involved, including local governments; private landowners; underserved and disadvantaged communities; scientists; business interests; commercial, recreational, and subsistence fishermen; agricultural groups; academic institutions; and nonprofits that have an interest in Gulf Coast restoration.

It is important to create a formal mechanism for public outreach and stakeholder engagement in this task force. All relevant options should be reviewed and evaluated to ensure formal, robust and comprehensive public engagement of the Gulf community throughout these restoration efforts.

It is also important that the task force reach out to federally-recognized tribes in the region to engage them in undertaking its responsibilities and determine an appropriate structure for tribal participation.

Task Force Structure

In order to conduct its work, the task force could be supported by technical teams that may be organized as described below. In organizing these technical teams, agencies should seek to involve staff people with relevant expertise who are located throughout their headquarters and regional offices. Effort should be made to ensure that all federal regional offices are engaged in supporting the task force.

If adopted, the teams described here should work together and are interdependent in fulfilling their responsibilities to support the task force.

- Regional Planning and Integration Technical Team: This team would be the technical group principally responsible for advising the task force on developing a Gulf of Mexico Regional Ecosystem Restoration Strategy, and for identifying gaps and planning priorities that will inform the efforts of both the task force and the Trustee Council.

  This team would advise the task force on integrating member agencies' activities into the Strategy. The team would consist of state and federal agency representatives with technical expertise in such vital areas as engineering, biology, oceanography, fisheries' management, environmental planning, and hydrology. This team could include members from the state advisory teams. Lastly, this team would be responsible for advising the task force on implementation of the Strategy and advising the Budget and Funding Coordination Team.
- Environmental Review Technical Coordination Team: This team would work with the Council on Environmental Quality (CEQ) to develop a strategy for environmental review under NEPA and state environmental review requirements, as well as other relevant authorities.

  NEPA requires federal agencies to consider environmental effects of their actions, including impacts to natural resources and cumulative effects, before deciding to proceed. NEPA also requires the analysis of social, cultural, and economic impacts. Moreover, NEPA provides stakeholders with an opportunity to be involved in the federal decision-making process, and ensures transparency with regard to complex issues and projects.

The team would consider several approaches, and then examine site-specific proposals through more focused NEPA analyses that tier from the overarching environmental review document. This team will work to coordinate its efforts with similar state laws and the environmental review approaches for the Deepwater Horizon oil spill NRDA Restoration Plan.

- Budget and Funding Coordination Team: This team would be responsible for working across agencies to identify and coordinate implementation of shared budget priorities for the region. As described in Section V, numerous potential funding sources exist that could fund restoration activities in the Gulf of Mexico region. Many of these programs are subject to annual appropriation, as well as to prioritization among various federal agencies' objectives and programs.
  This team would also assist in ensuring accountability with the Gulf Restoration Strategy and develop annual budget guidance in consultation with OMB. This team would synthesize input received from the Science Coordination Team and Environmental Review Technical Team on monitoring, research, and restoration project priorities.
- Science Coordination Team: The Science Coordination Team would provide support to the task force to ensure that decisions are based on practical and applied science. This includes coordinating the scientific and technical framework for ecosystem restoration and monitoring, including ensuring that restoration decisions are informed by science and that the scientific and technical work of restoration is properly coordinated so as to achieve the Strategy and ensure coordination between restoration science and monitoring and public health management and economic development needs.
  The team would be established to respond to needs identified by the task force and in a manner consistent with existing authorities such as the National Ocean Council, the National Science and Technology Council, and other bodies as appropriate, and establish effective mechanisms for engaging academic and other non-governmental scientists. This team could assist in development of review or selection criteria, performance measures and indicators to track progress in achieving restoration goals, or research priorities, including research into new methods that will provide innovative and sustainable solutions that address the challenges facing coastal communities. Deliverables would be provided to the task force as science advice. The team would coordinate with other teams of the task force to ensure that the principles of ecosystem based restoration and management, integrated ecosystem assessment, and socio-economic-environmental interdependencies are used as the strategy is implemented.

Additional technical teams could be established by the task force on either a permanent or temporary basis, as needed. A number of technical teams already exist in the Gulf region (many organized under GOMA or the Roadmap), all of which have state and federal agency staff participation. The task force could draw upon the expertise and capacity of these existing teams.

Task Force Near-Term Actions

The Ecosystem Restoration Task Force should be established and begin work as soon as possible. Upon its establishment, the members of the task force should develop and adopt a charter that outlines its functions and duties. This would include procedures for making decisions, resolving disputes, and addressing project-implementation needs. Development of the Strategy and the environmental review should be completed within a year and two years, respectively, of the establishment of the task force, and should build upon existing restoration planning and implementation efforts.

V. Funding Ecosystem Restoration

Accomplishing the vision for ecosystem restoration described in this Report will require an unprecedented level of cooperation and coordination among all levels of government and across the federal agencies. Moreover, as estimates to address existing coastal restoration needs make clear, the effort outlined in this report will require substantial investment. The recommendations in this chapter are related to ways that ecosystem restoration efforts might be improved within existing authorities. A number of federal statutory authorities currently exist that could be effectively coordinated with the work described here and use of these funds aligned to achieve an overall greater impact.

Federal agencies, in cooperation with the task force, should align implementation of these programs with a broader restoration vision. Moreover, the Gulf states have existing state sources of funding that have been highly effective in addressing coastal restoration needs. The task force should explore opportunities to align priorities between federal and state programs.

Finally, the effectiveness of restoration programs should be measured in terms of the Strategy, rather than on a program-by-program basis. No single program or authority will be sufficient on its own. It should be a principal responsibility of the task force to show how the combined efforts of all government programs are progressing toward restoration of this region as described in the Strategy, and to identify areas for improvement.

Potential funding sources to help achieve the vision of resilient and healthy Gulf of Mexico ecosystems include:

NRDA Damages

As explained above, under the Oil Pollution Act, responsible parties are liable to pay damages, including the cost of restoring, rehabilitating, replacing, or acquiring the equivalent of the injured natural resources and/or the ecological and human use services provided by the injured resources; the diminution in value of those natural resources pending restoration; and the reasonable costs of assessing those injuries. OPA requires that all sums recovered for restoration must be used to restore, rehabilitate, replace or acquire the equivalent of the injured or lost natural resources, including "emergency" restoration projects under certain regulatory conditions. Decisions regarding the use of NRDA funds can be made only by designated natural resource Trustees.

Coastal Impact Assistance Program (CIAP)

CIAP was created by Congress in 2005 to provide for ecosystem restoration for coastal states to help mitigate the impacts of oil and gas production in the Outer Continental Shelf beyond their shores. The Secretary of the Interior is authorized to distribute to producing states and coastal political subdivisions $250 million for each of the fiscal years 2007 through 2010. While existing CIAP balances could be a source of funding for long-term restoration and recovery, the Gulf states have reported that they have encountered obstacles that have greatly limited their abilities to access these funds. Identifying and addressing any roadblocks in a timely fashion would further the goal of providing funding immediately to implement habitat protection and restoration projects. States could prioritize their use of CIAP funding for the highest priority projects in coordination with the task force and consistent with the final Gulf of Mexico Regional Restoration Strategy.

Coastal Wetlands Planning, Protection, and Restoration Act (CWPPRA)

The Coastal Wetlands Planning, Protection and Restoration Act (CWPPRA) is federal legislation enacted in 1990 that is designed to identify, prepare, and fund construction of coastal wetlands restoration projects. Since its inception, 145 coastal restoration or protection projects have been built, benefiting over 110,000 acres in Louisiana.[24] The annual budget for CWPPRA funded restoration has varied through the nearly 20-year life span of the Act. The budget, dependent on revenues from taxes on fishing equipment, import duties, and small engine motorboat fuels, has ranged between approximately $30 million per year to nearly $80 million per year.[25] The funded Louisiana projects provide for the long-term conservation of wetlands and dependent fish and wildlife populations with cost-effective plans for creating, restoring, protecting, or enhancing coastal wetlands.

Discretionary Appropriations

Another potential funding source may be found in the discretionary requests proposed under the various federal agencies' existing authorities.. While this approach has its challenges, it also has the advantage of being straightforward, and would provide certainty about available funding. It would not be dependent on unknown future penalty or settlement figures, or similar future anticipated receipts of unknown size and timing.

VI. Conclusion

The Deepwater Horizon oil spill has had tragic consequences for Gulf ecosystems and the Gulf economy, exacerbating existing restoration challenges and creating new ones in a region that has struggled with a legacy of environmental degradation. At the same time, it provides a tremendous opportunity to focus the nation's attention on addressing the latest challenge to this tremendously productive and diverse ecosystem.

This report outlines a path forward to maximize the opportunity for resilient and healthy, Gulf of Mexico ecosystems that support the diverse economies, communities, and cultures of the region. With statutory authority, the new governance model would provide a robust mechanism in which federal and state agencies can work together to coordinate their efforts, engage stakeholders, and resolve the policy and process obstacles to implementation that will help align and move forward restoration plans and projects. In a region that has long suffered from no shortage of plans but rather a lack of coordinated prioritization and implementation of those federal, state, and local plans, this report offers a strategic framework that can align the myriad regional interests and operations under a single structure so that the most pressing Gulf-wide ecosystem challenges can be addressed comprehensively and effectively.

Restoration of the Gulf of Mexico region will take sustained leadership and engagement by multiple stakeholders, and a significant commitment of resources over many years. This is a partnership and a collaboration that requires participation by (and investment from) all parties –federal, state, tribal, and local governments, non-profit organizations; the private sector; academics; and scientists. Together, the region and the nation can build a healthy and resilient Gulf Coast that can meet both this and future challenges.

Health and Human Services Recovery

I. Introduction

The Deepwater Horizon oil spill has had a profound impact on the lives and livelihoods of Gulf Coast residents; the region's environment, wildlife, and seafood; and the economy of all five states that border the Gulf. The sudden and emerging environmental disaster created by the spill has caused special health and economic needs for Gulf residents that must be addressed.[26] While all of the impacts are still unknown, effects of the spill are expected to continue for years to come.

Health and human services address a wide array of critical human needs. Strategic planning for health and human services recovery must be informed by the principle that federal efforts are designed to support locally driven priorities and needs in conjunction with affected states and communities. A strategic approach must also recognize that many interrelated systems are needed to support individual and community health, resilience, and well-being. These include the medical care and public health systems, behavioral health care, social services, and prevention. They also include elements essential to health such as equal access to care, employment and financial stability, functioning communities, and a safe environment.

In order for affected Gulf Coast communities to regain their momentum, it is essential that a long-term commitment be made to restoring the environment, seafood supply, and health of the region's population. This chapter details the health and human services challenges posed by the oil spill, reviews current activities, and proposes strategies to assist the people of the Gulf Coast in supporting themselves as they recover from this tragedy. The response to these challenges and the methods to recover from them span a number of domains. These include ***surveillance***, ***behavioral health*** prevention and treatment, the delivery of ***medical services***, access to ***human services***, ensuring ***seafood safety***, and cooperative and coordinated ***research***.[27]

II. Background Information

A. Pre-Existing Health and Human Services Challenges in the Region

Since Hurricanes Katrina and Rita, the Gulf Coast has experienced continuous overlapping cycles of emergency preparedness, response, and recovery. Health care, the functional needs of at- risk individuals, behavioral health, social services, and other service providers were neither exempt from these effects, nor from the destruction of homes and livelihoods. Because they had the resources to do so, many professionals left the area ahead of the storms, never to return. Many facilities were destroyed and have yet to be rebuilt, and some service providers continue to operate out of temporary structures or at locations with lower quality of service due to staffing challenges.[28]

Among the greatest existing health and human services recovery challenges in this region are:

- Loss of the professional behavioral health, health care, and social services workforce and the resulting impact on the recovery of the local economy.
- Acute and chronic health conditions in a population – much of which is disadvantaged – with insufficient access to healthcare.
- Reduced access to health care due to fewer delivery sites for services, with many sites still closed or operating in temporary, often smaller facilities.
- Increased demand for health, behavioral health, and community and social support services that are already stressed by decreased donations, lower revenue, and less functional facilities.

This damaged and fragile infrastructure was still moving toward full recovery when the oil spill occurred. All of this has produced challenges for state and local governments throughout the region, who now struggle to address these needs in an environment of reduced resources and personnel.

B. How the Deepwater Horizon Oil Spill Further Impacts the Region

Federal and state agency scientists are working to better understand and quantify the impact on health and human service needs caused by the spill. Although studies exist on the health impacts of other oil spills, the scale of the Deepwater Horizon oil spill makes it difficult to fully predict, with specificity, either the potential short- and long-term effects to both workers and residents of the region, or the needs that will result.[29] While coastal residents and tourists were urged to avoid close contact with oil, response workers may have been exposed to many different chemical, physical, and biological hazards depending on the type and stage of response and the workers' specific tasks. In addition, there are segments of the population along the Gulf Coast that have chronic health issues and lack access to adequate health care.[30] Should surveillance and research activities discover a connection between exposure and negative health effects, this could further affect health care access in the region.

While further research is needed to determine the physical health implications of the oil spill, behavioral health issues have already become a primary health concern in the immediate response. Evidence from past disasters, including the *Exxon Valdez* oil spill and Hurricane

Katrina, show that behavioral health needs, and related issues such as domestic violence and family dysfunction, are likely to increase over time and become a greater burden on health care, public health, and social service systems during the recovery period.[31] This specific type of disaster can create a loss of social and economic support, causing significant anger and resentment, which can impact the human stress reaction and contribute to a host of medical conditions.[32,33] These stress reactions are real and can lead to specific mental and substance abuse patterns that may exacerbate or lead to various medical conditions. Such conditions include diabetes, hypertension, heart disease, and other illnesses. They may also profoundly impact a resident's life span and quality of life.[34]

Because of its impact upon the economic well-being of the Gulf Coast, the oil spill has important human services implications. If jobs are lost in key regional industries, aggravating already difficult circumstances for many Gulf Coast residents, large segments of the Gulf Coast population will experience an intense need for economic and social support. An increased demand for a range of federal, state, and local programs, including those providing support for income, home maintenance, and stress-related family challenges is already occurring.[35,36]

Seafood safety concerns further impact the region's health and economic climate. The crude oil in the Gulf waters as a result of the spill is a combination of many different chemicals, a number of which are harmful to people if inhaled, absorbed through the skin, or ingested in contaminated food or water. In addition, chemical dispersants were used both on the surface and at the well head to facilitate the oil's degradation. Furthermore, long-term accumulations of contaminants such as heavy metals, present in crude oil, will be monitored for years. Therefore, sampling, analysis, and surveillance of various types of seafood from affected and adjacent areas will continue to ensure that harvested seafood is untainted by oil spill-related contaminants.

C. Initial and Current Response to the Oil Spill

At the time of the spill, there were a number of entities working on Gulf Coast health and human services recovery activities as a result of damage inflicted by Hurricanes Katrina and Rita. These included federal agencies, state and local governments, federally and state-recognized tribes, private corporations, non-governmental organizations, academic and scientific organizations, other affected stakeholders, and the general public. The economic and environmental uncertainty as oil continued to flow from the wellhead and as the beaches, marshes, and waters became visibly polluted, presented a significant challenge to these efforts. As a result of the spill, federal departments and agencies, state and local governments, and others have responded to the immediate health and human services needs of the Gulf Coast states. But many less tangible needs continue to evolve as the recovery phase of operations begins. The following summaries provide a non-exhaustive overview of some of the major health and human services response efforts underway on the Gulf Coast.

Surveillance

Since the oil spill began, there have been concerns about the extent to which related hazards, such as physical and chemical exposures and social and economic disruptions, may impact human health. Using state-based and national surveillance systems and BP injury and illness data, Gulf Coast states and the HHS Centers for Disease Control and Prevention (CDC) are tracking short-term health effects potentially related to the oil spill in order to alert public health officials to trends that can be addressed or to trends that require further investigation.

At the request of public health officials in Alabama, CDC helped state and local officials conduct Community Assessment for Public Health Response (CASPER) surveys in two counties. A CASPER rapidly measures the health status and basic needs of the affected community and provides objective information to public health practitioners and emergency management officials in order to direct assets. Conducting a CASPER provides state and local health officials with a more complete assessment of public health needs, including behavioral health needs. Additionally, EPA has monitored air and water for human health impacts.

Behavioral Health

To date, officials from Gulf Coast states report an increase in the need for behavioral health services. They also express concern about behavioral health issues related to job loss, as well as a perceived loss of the Gulf's distinct culture and way of life. As a result, community organizations, state officials, and traditional providers have been conducting outreach, training, and other services for those who exhibit fear, sadness, irritability, and loss of hope. At the federal level, HHS has been directing attention and resources to address the behavioral health issues arising from the oil spill. It has also sought to ensure that at-risk individuals and populations such as children, the elderly, and limited English proficient speakers, are considered in the Gulf Coast region's response efforts.[37]

HHS is engaged primarily in support of state and local efforts to assess and meet the behavioral health needs of those responding to the spill. Efforts focus both on prevention of long-term behavioral health problems, and provision of services for those who need them. To aid state efforts, the HHS Substance Abuse and Mental Health Services Administration (SAM HSA) has published and widely disseminated six "Tip Sheets" for coping with the disaster, including one targeting teachers and school administrators.[38] These tip sheets have been translated into multiple languages. In addition, HHS is developing a single number to access Gulf Coast disaster distress hotline counseling services. SAMHSA also regularly convenes Disaster Behavioral Health Coordinators in the Gulf Coast states to facilitate the provision of technical assistance, assess the impact of this event in each state, create opportunities for states to share information, and allow the Department to maintain an understanding of the current situation. The Department is also in communication with NGOs, such as the American Red Cross and Catholic Charities, that are active in the affected areas and are engaged in addressing the functional needs of at-risk individuals and behavioral health concerns.

Medical Services

In response to the oil spill, HHS activated the National Disaster Medical System as a precaution, should Gulf Coast states need additional medical support. In addition, an HHS mobile medical unit – located in Venice, Louisiana, and staffed by an HHS medical team comprised of a doctor, two nurses, two emergency medical technician-paramedics and a pharmacist – has supported the local medical community by triaging responders who are concerned about the health effects of the oil spill. Patients, who required medical attention beyond the basic care available through the mobile unit, were referred to local healthcare providers and hospitals.

Human Services

Federal human services program offices have been leaning forward to anticipate, prepare for, and respond to potential changes in service needs created by the oil spill. Many have been communicating directly with the Gulf Coast states about any immediate burdens created by the oil spill. Other program offices are seeking information directly from grantees in affected states, from Community Action Agencies or Community Action Programs. The H HS Administration for Children and Families (ACF) has observed a 13% increase in monthly calls to the National Domestic Violence Hotline from the Gulf Coast states between April and June 2010, including a 21% increase in calls from Louisiana.[39] ACF continues to track usage of the National Hotline. In addition, because state and local domestic violence programs respond to the bulk of crisis calls from victims of domestic violence, Gulf Coast state domestic violence coalitions are monitoring crisis calls to state hotlines and local domestic shelter usage. ACF's Family Violence Prevention and Services Program is also working with state domestic violence coalitions in the Gulf Coast states to identify issues of concern and respond to technical assistance needs.

Seafood Safety

The federal government led a comprehensive program to protect seafood safety by ensuring that seafood caught in the Gulf of Mexico is free from contamination as a result of the oil spill. This program is important not only for consumers who need to know their seafood is safe, but also for the fisheries industry, which needs to be able to sell its products with confidence. This program has been instituted in full cooperation with the states.

The comprehensive program was led by the HHS Food and Drug Administration (FDA) and the Department of Commerce's National Oceanic and Atmospheric Administration (NOAA), and included the Environmental Protection Agency (EPA), other federal agencies, and state authorities in the Gulf Coast region. The federal agencies have been taking a multi-pronged approach to ensure that marketed seafood from the Gulf of Mexico is safe. These measures include the precautionary closure of fisheries, surveillance and testing of seafood products, and inspections of primary seafood processing firms for compliance with FDA's Hazard Analysis and Critical Control Point regulations. Federal agencies, working collaboratively with the Gulf Coast states, implemented a protocol to determine when closed federal and state harvest waters can be reopened. This protocol involves sensory testing for polycyclic aromatic hydrocarbon (PAH) components of the oil and dispersant along with chemical-based testing for PAH as a confirmatory measure. Going forward, it will be important to continue to strengthen coordination among state laboratory systems, federal agencies, and laboratory networks and enhance the ability to exchange data, results, and analytical methods.

Research

Numerous workshops and panels have been held to define priorities for health issues related to the Gulf oil spill and to discuss the following matters: the potential health effects of the oil spill, the short- and long-term effects of oil exposure on human health, current federal and state monitoring efforts, potential research modalities, and the characteristics of effective surveillance and monitoring systems.

Recommendations from previous spills, particularly from an Institute of Medicine workshop, are for prospective, long-term epidemiological studies to assess potential health effects associated with clean-up activities following an oil spill. In addition, both short- and long-term laboratory and human health studies involving local communities are necessary to assess the wide range of potential health consequences, and to develop the evidence base to inform present and future recovery and health care services. It will be essential that the federal government work in partnership across agencies and with local communities on the Gulf Coast in the planning and implementation of health research activities.

Future studies should address a range of potential health outcomes, such as cancer and effects on respiratory, reproductive, immune, and nervous (including neurobehavioral) systems; skin, renal function, and behavioral health effects. New research to establish sustainable models for delivering mental health care – when the number of people in need exceeds capacity by orders of magnitude – is needed. Such research has relevance for not only domestic preparedness and response, but also for non-emergency contexts where mental health infrastructure is limited (i.e., rural mental health), and for the United States' role in global health crises. The human health assessments should be coordinated with the ecosystem restoration science and monitoring efforts. It is critical to the sustained resilience of coastal communities that these efforts enhance understanding of the connection between ocean health and human health.

Soon after the spill occurred, CDC's National Institute for Occupational Health and Safety (NIOSH) conducted Health Hazard Evaluations (HHE) and surveillance of reported illnesses among workers involved in the oil spill response. As each HHE is completed, the findings are reported to BP and then posted to the NIOSH website.[40] NIOSH also collects and analyzes the acute injury and illness data that BP reported to the Department of Labor's Occupational Safety and Health Administration (OSHA). This NIOSH work is important for an understanding of the physical and mental health risks associated with Gulf Coast oil spill response work. NIOSH, in support of future research efforts, has compiled a roster of workers who participated in the response, to create a record of work and mechanisms to contact them about their spill-related symptoms of illness or injury.

More than 54,500 responders have been added to the roster, including BP-trained workers, volunteers, vessel of opportunity operators, and federal workers. NIOSH compiles the roster through a voluntary system at worker staging areas, during worker training, and through a secure web site. NIOSH is currently engaged in acute animal toxicity studies of both dispersants and crude oil from this well. Following the evaluation of toxicity studies results, NIOSH may consider other activities in follow-up. NIOSH's toxicity studies are included in a suite of studies planned by the National Toxicology Program of which NIOSH is a member.

Additionally, CDC is reviewing EPA sampling data to determine whether these exposures to oil, oil constituents, or dispersants might have caused human health effects. The data includes sampling results for air, water, soil/sediment, and waste oil samples. If the surveillance or monitoring systems identify potential health risks or groups of people with symptoms, state and local public health officials are able to follow up as needed to investigate whether there is an association between the symptoms and the oil spill.

D. The Economic Nonprofit Health and Human Services Nexus

Thriving, resilient communities are the result of strong private, public, nonprofit, and nongovernmental organizations. Nowhere in the nation does government—local, state, or federal—provide all the resources required for successful communities. Sustainable employment opportunities across a range of occupations and sectors generate income for those employed; taxes provide support for various levels of governance; and donations to nonprofits, community and faith-based organizations, voluntary groups, and other charities provide a health and social safety net for many. Any threat to the delicate balance of a community's economic well-being can disrupt that safety net. Short-term disruptions often bring communities together, with people giving time, goods, or money to bridge a crisis. When communities experience longer-term disruptions, such as widespread job loss concentrated in a geographic area, it can significantly disrupt quality of life as residents are forced to abandon their livelihoods, drive long distances to jobs, or lose hope of any positive turnaround.

Oil production, fishing, and tourism are mainstays of the Gulf Coast economy. The oil spill has reverberated throughout all three sectors, with corresponding effects on associated businesses. Additionally, donations to nonprofits, community and faith-based organizations, voluntary organizations, and other charities have been significantly reduced. In turn, these stresses have provoked an increase in anxiety, hopelessness, and despair, just as the decline in resources has limited the capacity of the nonprofit sector to respond.

III. Principles for Health and Human Services Recovery

In order to achieve true recovery, a comprehensive framework is required to move beyond the cycle of continual response to recurrent disasters and provide the Gulf Coast with the infrastructure, information, and services it needs to recover. The plan for the recovery of health and human services following the oil spill is consistent with the principles of recovery and the vision of health security as set forth by the National Health Security Strategy.[41] Specifically, the proposed approach is built on a foundation of community resilience: healthy individuals, families, and communities with access to health care and with the knowledge and resources to be able to care for themselves and each other in both routine and emergency situations.

The plan addresses a set of interrelated health and human service needs and describes how the federal government can support the primary efforts being carried out at the state and local levels to optimize conditions for recovery of health and human services. In the context of a complex disaster like the Deepwater Horizon oil spill, it is critically important to ensure that ongoing ***surveillance*** and scientific ***research*** inform the recovery agenda and that priorities include ***behavioral health*** prevention and treatment, the delivery of ***medical services***, and access to ***human services***. Likewise, ensuring ***seafood safety*** protects not only individual health, but promotes economic recovery by mitigating potential individual, family, and community dysfunction. All of these domains must be based on empirical evidence made possible by ongoing surveillance efforts and targeted scientific research. As shown in Figure

1 below, in this framework each element informs the others and promotes continual refinement of planning and recovery activities to meet the recovery needs of this disaster and build a foundation for addressing new threats experienced by the Gulf Coast.

Goal 1: Design federal efforts to support locally driven priorities and needs with extensive collaboration with affected states and communities.

Goal 2: Initiate surveillance and monitoring of the health status and well-being of the impacted population early, with continued surveillance and monitoring activities for a number of years, as needed.

Goal 3: Ensure access to medical and behavioral health care and human services support. As surveillance uncovers potential health problems, connect findings to capacity for appropriate medical and behavioral health care and provision of human services programs.

Goal 4: Continue to ensure that the seafood on the market is safe, which is central to both public health and the economic recovery of the Gulf Coast.

Goal 5: Continue to develop the health, behavioral health, and human services evidence base in order to inform recovery and health care resource needs for the Deepwater Horizon oil spill and future disasters.

Vision: The restoration and resilience of the health and well-being of affected people and communities.

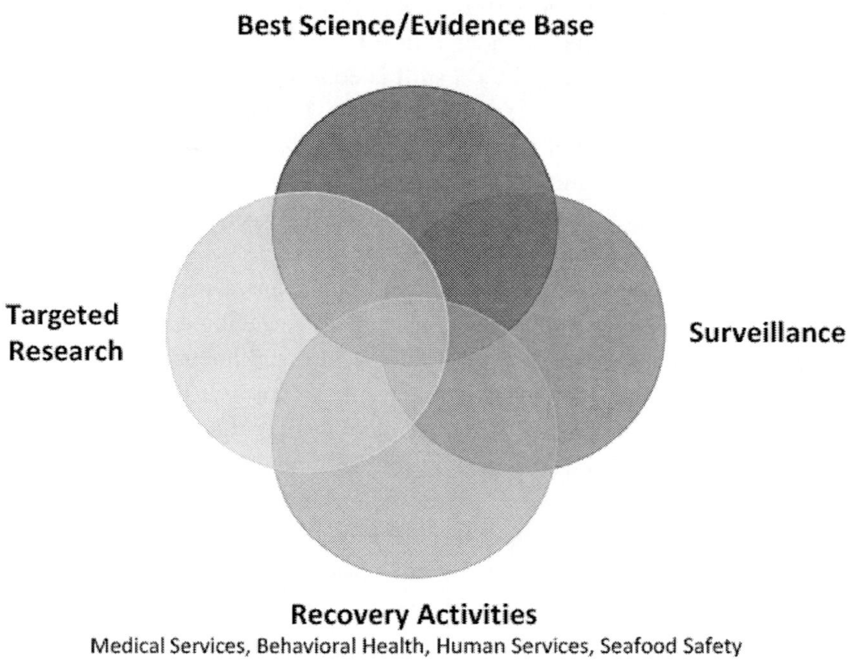

Figure 1. Health and Human Services Recovery Framework.

IV. Recovery Strategy

Much has been learned from prior planning efforts in the region about the critical need to promote cooperation and coordination among various federal, state, and local entities in order to align, refine, and adapt recovery plans and projects. The evolving nature of the oil spill requires a particularly active collaborative process governed by community needs and carried out through local-state-federal and public-private partnerships. The following section describes seven categories of recovery challenges, recovery activities already underway, and additional recovery needs. Some of the recovery activities currently underway were begun with initial funding support from BP and/or the federal government; however, none of these can be fully implemented as described without additional funding. Further, all other recovery needs listed below will require new funding before implementation. That is why it is critical that the responsible parties for the spill also do their part for the recovery of the Gulf Coast.

A. Surveillance

For public heath, behavioral health, and human services recovery efforts alike, effective systems of surveillance are critical to identify needs, target response actions, monitor conditions, and provide ongoing feedback on the efficacy of interventions. Yet, public health, behavioral health, and human services response efforts are all challenged by limitations in existing data, surveillance assets, and collection methods.

Studies of previous oil spills provide some basis for identifying and mitigating the human health effects of oil exposure. However, the existing data are insufficient to fully understand and predict the overall public health impact of spill hazards on the health of workers, coastal residents, and at-risk populations such as pregnant women, children, and the elderly. The surveillance systems currently employed are designed to look for health trends in different ways and to alert health authorities to signs of possible public health impacts due to exposure to the oil or chemical mixtures associated with the oil spill. Yet, the systems will need amplification. While behavioral health issues–such as psychological and emotional distress, depression and substance abuse – and related problems, such as domestic violence, have already surfaced as primary health concerns, few of the current surveillance systems focus on such issues.

By fully utilizing and further developing public health surveillance systems, particularly systems that focus on behavioral health, the public health community will be able to gain a more complete picture and better understanding of the health issues related to the spill. It will also be able to share this information with the public. Enhanced surveillance coupled with targeted research will contribute to evidence-based information regarding the health effects of the oil spill that will enhance the ability of those in leadership positions to make informed decisions about mitigation activities and strategies.

In human services surveillance, the sudden and rapidly evolving economic changes in the wake of the recent oil spill require greater coordination of effort to ensure human services are delivered where most needed. Combining improved surveillance with existing data would allow HHS to more immediately target and enhance services in critical need areas on the Gulf Coast, as well as inform recovery for future disasters.

Recommended Recovery Actions

Needs assessments are being funded on the Gulf Coast; however, all of the Gulf Coast states need both technical assistance and additional funding to expand current capacity for data collection and surveillance, in order to allow for long-term needs assessment and planning activities.

The following surveillance activities are critical to Gulf Coast recovery and community resiliency. Some have initial financial support from BP, but as explained above, few can be fully implemented without additional funding. These activities should be coordinated with the states, they should enhance ongoing surveillance activities, and they should subsequently inform recovery activities related to prevention and services:

- Conduct a telephone survey for the assessment of behavioral health and associated health impacts in the Gulf area. In partnership with state public health and mental health departments and SAM HSA, CDC will design and conduct a telephone survey in the affected coastal regions of the Gulf Coast states to monitor mental health status, including measures of psychological stress, anxiety, depression, potential stress-associated physical health effects, and other indicators. Some questions included in the survey will be pulled from the ongoing Behavioral Risk Factor Surveillance System (BRFSS), from which preliminary data show increases in several indicators of emotional stress.[42]
- Assess behavioral health needs via a new data collection tool developed and piloted for the Federal Emergency Management Agency Crisis Counseling Assistance and Training Program.
  These forms would be used to gather demographic information, risk factors, signs and symptoms, and delineations of services provided. Available SAMHSA Emergency Response Grant (SERG) funding will support state use of the forms and the infrastructure for regular information reporting to SAMHSA.
- Use the National Survey on Drug Use and Health to compare the prevalence of substance use and mental health problems before, during, and after the oil spill. An over-sampling of the affected areas in the Gulf Coast region for up to two years would permit long-term surveillance of changes in the region's behavioral health status. This work would allow HHS to better understand both the long-term behavioral health needs and the impact of behavioral health interventions on recovery.
- Develop a comprehensive human services data tool that would consolidate existing data and provide real-time human services information, as discussed in the human services section of this chapter. Additional funding is needed to develop this tool, which would contain trend data at the city, county, and state levels for numerous contextual indicators. These indicators would include the number of residents by age, education, employment status, occupation, income, residence type, family structure, language preference, and home internet access. In addition, participation rates in Temporary Assistance for Needy Families (TANF), Supplemental Nutrition Assistance Program (SNAP), Head Start, and other key programs, as well as home foreclosure rates, would be included. HHS would work with partners across state and

federal governments to create and update this tool, which is critical to understanding local contexts and progress in this disaster, as well as in future disasters across the country.

- Use the National Intimate Partner and Sexual Violence Surveillance System (NISVSS) to collect enhanced data on the impact of the crisis on domestic violence. Additional funds would be needed to expand the existing sample population to get a better understanding of the impacts of the oil spill on domestic violence rates.
- Support the operations of the five impacted Poison Control Centers (PCC) supported by the HHS Health Resources and Services Administration, including fielding calls from the public. Such support will make PCCs whole for expenditures for the additional call volume both to date and going forward, with the expectation that there will continue to be a high volume of calls over the next six months.
- Continue to monitor Medicaid and Medicare health care claims and program enrollment to determine whether at risk populations have access to care. This is being done through the use of an industry code for oil spill injuries or illness claims for hospitals and institutional providers and a Healthcare Common Procedure Coding System "CS" modifier for physician and other practitioner claims. These codes will assist in identifying health care claims due to oil spill illnesses or injuries and may be useful for program integrity, research purposes, and improving access to care.

B. Behavioral Health

The 20.5 million people living in the Gulf Coast region, and those dependent on the area for their livelihoods, may be adversely affected by the oil spill disaster for years to come. Significant among the consequences are the behavioral health effects of this disaster. The consequences likely will be especially prevalent for those who are directly impacted, but likely will also impact those "upstream" who provide care and services for individuals and families, and those whose livelihoods are dependent on the efforts of those individuals and families. According to reports from state behavioral health officials, some in the affected population are already exhibiting early signs of developing substance abuse and dependence, psychiatric disorders, suicidal risk, and familial breakdown, including domestic violence and child abuse. Addressing these needs now and over time for the residents and clean-up workers is crucial to the overall response to behavioral health concerns in the Gulf Coast.

The disaster represents a loss of social and economic support for people, causing anger and resentment toward each other and outsiders.[43] BP claims reports show that in the first 60 days of the disaster alone, one-third of all health claims associated with this oil spill have been for anxiety and stress. If these conditions are left untreated, it is expected that the mental health condition of these individuals will worsen.

The three most prevalent psychiatric diagnoses experienced following the *Exxon Valdez* oil spill were anxiety, depression, and post-traumatic stress disorder (PTSD). Research indicates that individuals closer to the affected area were 3.6 times more likely to experience generalized anxiety disorder and 2.9 times more likely to develop PTSD. Fishermen from local communities reported the most psychological distress. Additionally, the economic cost associated with other health conditions, domestic violence, criminal justice, education, and lost productivity will increase if mental health issues are ignored.[44]

People suffer adverse physical health effects from stress and stress reactions. Stress is linked to several leading causes of death, including heart disease, cancer, lung ailments, unintentional injuries, cirrhosis of the liver, and suicide. Behavioral health conditions can manifest themselves physically as headaches, gastrointestinal disorders, insomnia, inability to eat or eating too much, increased use of alcohol, or use of substances of abuse, including prescription drugs. The link between behavioral health and physical health is one that must be understood in the context of this disaster.

These physical and behavioral reactions to stress are associated with community disruptions. These can include increased arrests for driving under the influence and an increase in domestic violence, assault, child abuse, and child welfare cases.[45] Effects also can include increased difficulty in early learning centers and schools by children affected by the spill; an increase in bullying, truancy, and inappropriate behavior, which often leads to more emotional distress on teachers and other children; reduced educational performance; and the possibility of other severe behaviors.

Gulf communities have shown incredible resilience and strong cultural and community commitment to recovery. However, state behavioral health professionals report that their systems are chronically stretched to meet the needs of their citizens. Communities and state officials need help to fund and conduct outreach efforts, psychological first aid, brief interventions, and peer support. They also need help in providing timely and culturally responsive treatment and services for disorders that emerge.

Recommended Recovery Actions

In order to meet the expanding needs of Gulf Coast residents, it will be important to ensure that behavioral health services are informed by both existing and enhanced surveillance efforts and an evidence base described by current and developing research. Gulf Coast state officials report the need for assistance in the provision of comprehensive services to address the behavioral health issues caused by the oil spill. States indicate needing approximately $97 million for the first year and over $400 million for five years to provide interventions to prevent behavioral health disorders as well as to provide both crisis counseling and treatment services. BP recently announced its intention to contribute $42 million toward these state-identified needs and another $10 million to federal needs.

From prior disaster research, service needs within the general population can be estimated as follows for the first five years:[46]

- 15 - 30% of the estimated population at risk will need no help at all.
- 30 - 60% will respond to the following:
  - Community supports and service
  - Public messaging
  - Outreach and information dissemination
- 25 - 35% of the population at risk will need increased levels of clinical care:
  - Screening and assessment for current or emerging substance abuse and mental health concerns
  - Evidence-based counseling by trained professionals or para-professionals, by telephone, telehealth capacity, or face-to-face, in short and infrequent interactions, individually or in groups (e.g., with fishermen or oil spill workers as they come in from their shifts)

- Brief/short-term interventions in non-specialty settings (for example, primary care or human services settings or mobile medical units)
- Referral and assistance accessing professional mental health and substance abuse treatment and services utilizing private insurance and public assistance where possible (although coverage for gaps in insurance will be significant in these high- uninsured states[47])
- Inpatient, outpatient, and medication treatment services for those with diagnosed ongoing disorders

The federal government is committed to continued work with the Gulf Coast states and support their implementation of recovery activities. Federal actions can be placed into elements that would extend state capability and be usable by all within the regions. HHS has identified the following behavioral health activities, that have received first year funds from BP, as critical to recovery. There may be unknown additional needs beyond these activities:

- Develop enhanced media outreach materials. It is critical in this disaster to provide information to the public informing them that strong emotions are to be expected; that it is important to stay connected with friends and family; that families and neighbors should take care of themselves, their children, and each other; that peer or professional help may be needed to prevent or treat behavioral health issues and the circumstances in which they occur; and where to find help when it is needed.
- Establish a toll-free behavioral health hotline number explicitly for disaster-affected individuals and families that will use the National Suicide Prevention Lifeline network's electronic telephone system to connect callers in distress to the nearest Lifeline member crisis center in the Gulf Coast area. This service will assist states and local communities in meeting the many complex behavioral health needs resulting from the Deepwater Horizon oil spill.

C. Medical Services

The healthcare systems within the Gulf Coast region previously impacted by hurricanes of the past have continued to struggle during recovery. Rural populations in coastal as well as inland areas are already heavily burdened with chronic illness. They have also faced daunting challenges to accessing medical care as primary care practitioners and networks have left the affected area.[48] Poor access to medical care not only challenges the continued medical care of the chronically ill and those most susceptible to exposure-related illness, but also the ability of surveillance to identify emerging illness and provide the required level of medical care. Recovery strategies must be based on science and the best current surveillance available to improve access to medical care, prevent and mitigate illness and injury to at-risk individuals, and respond to any emerging exposure-related disease.

Monitoring responders and community members exposed to contaminants from the Gulf Coast oil spill will be important to identifying any emerging associated disease burdens. The known toxic effects of crude oil exposure make oil spill responders working near or at sites of oil plumes, burn-off locations, and clean-up operations, an obvious target for short- and long-term surveillance for health effects. Other potential exposures exist due to dispersants and mixtures of associated chemicals and crude oil in various stages of degradation. The routes of exposure to the dispersants and complex mixtures may occur through inhalation, skin contact, swallowing, or through ingestion of contaminated food.[49] Environmental sampling data collected by EPA suggests that the likelihood of exposure of community members to oil or dispersant contaminants is low. However, many of the contracted responder population involved in the response come from rural and remote coastal communities.

Existing studies of workers and residents near oil spills report short-term health effects from exposures to crude oil and oil mixtures, including immune function effects, respiratory, skin, and eye irritation.[50] Effects were found to be greatest in clean-up workers and commercial and subsistence fishermen involved in clean-up efforts. In one study following the *Prestige* oil spill, fishermen involved in the clean-up were found to have increased levels of lower respiratory tract symptoms more than a year after the accident, and risk increased with increasing work exposure to oil and oil mixtures. Several studies indicate that women may be particularly vulnerable to the effects of exposure to oil and clean-up activities, although few have assessed adverse effects on maternal, child, and reproductive health.[51]

A few studies involved collection of environmental samples of air, water, or dust to assess exposure levels in relation to reported health effects. A smaller number of studies involved collection of human biospecimens, such as urine or blood, to assess body burden, renal and endocrine function, and DNA damage. Only two studies on the *Braer* and *Prestige* oil spills reported evidence of DNA damage and alterations in hormonal status related to exposure; and in both, the analysis involved only a small subset of the exposed population.

Compounding the uncertainty of assessing any short-term and long-term health effects from crude oil spill exposures are the various degrees of degraded oil exposures, new and poorly understood widespread exposure to dispersant chemicals, and the mixture of these substances. Despite the large number of spill disasters and affected populations in the United States and globally, we know relatively little about the short- and long-term health consequences that could help direct where limited resources are allocated for environmental clean-up, medical and social services, and surveillance. Potential long-term health effects could include cancer and adverse effects on immune, cardiovascular, respiratory, neurologic, reproductive, dermal, and neurobehavioral and neurocognitive function.[52,53]

The short- and long-term health effects that are identified by ongoing surveillance and toxicological studies will thus place additional burdens on an already impacted healthcare system in the region. Responders will need continued care for acute injuries and illnesses while cleanup operations continue. While BP has provided onsite management of acutely ill or injured workers, the recovery plan needs to consider workers who wish to seek care outside of BP established care stations, those who seek care after being released from employment with BP or related oil-response contractors, and those who need specialty care management during the recovery phase. Private practitioners will need to support ongoing surveillance efforts by reporting oil-related conditions to existing and future state and federal surveillance systems. Efficient and timely mechanisms for claim filing and reimbursement related to oil exposure also need to be in place to manage claims that occur both in the short- and long-term.

Health-related claims will need continued access to the claims process if long-term health effects are identified and any emerging healthcare burden is realized. The improved access to care necessary to address uninsured and underserved who have oil-related health effects will depend on additional funding. Any additional oil exposure-related burdens realized by emergency departments, definitive care facilities, and specialty care providers will also require reliable, efficient, and timely reimbursement.

Recommended Recovery Actions

Providing improved access to local healthcare systems that are recovering from previous hurricanes will be challenging. HHS has identified the following recovery activity needs. Recovery needs may evolve as we learn more from surveillance and research efforts.

- *Increase access points to medical care.*
  On August 9, 2010, HHS Secretary Kathleen Sebelius announced the availability of up to $250 million in grants to support approximately 350 New Access Points for the delivery of primary health care services for underserved and vulnerable populations under the Health Center Program in FY 2011.[54] The funds, made available by the Affordable Care Act, will be awarded by HHS's Health Resources and Services Administration (HRSA) through a competitive grant process. A new access point is a new full-time service delivery site that provides comprehensive primary and preventive health care services. New access points have the potential to improve the health status and decrease health disparities of the medically underserved populations to be served. Competitive applications will demonstrate a high level of need in the community, present a sound proposal to meet this need by ensuring the availability and accessibility of essential primary and preventative health services, and show that the organization is ready to quickly put its proposal to work. Organizations eligible to compete include public or nonprofit private entities, including tribal, faith-based, and community-based organizations that meet health center funding requirements. Current HRSA grantees may apply, as well as organizations that are applying for the first time. In addition, unsuccessful applicants from prior year competitive grant announcements may also apply.
  Locally managed networks of mobile health clinics can provide additional access and outreach to healthcare in areas with poor access, low population densities, and/or uninsured populations.[55] Support of existing networks should provide expansion into underserved areas, assist in building sustained access to primary care, improve surveillance of oil and dispersant- related disease, and assist in addressing any related surge in medical demand.
  Community organizations that meet Federally Qualified Health Center (FQHC) requirements but do not receive a competitive grant can apply to the FQHC "Look-Alike" Program. HRSA may recommend to HHS' Centers for Medicare and Medicaid Services (CMS) that entities meeting qualifying criteria be designated as FQHC Look-Alikes, which allows the entities to receive many FQHC benefits, including enhanced Medicare and Medicaid reimbursement. HRSA and CMS work closely together to administer the FQHC Look-Alike Program, which helps expand services to those who lack access to them and to retaining health care professionals.
- *Improve access to care at existing local healthcare systems.*
  Federally qualified health centers funded by HRSA, rural health clinics, and other sites that care for low-income and uninsured people can become National Health Service Corps (NHSC) approved sites. At such sites, dentists, dental hygienists, primary care physicians, nurse practitioners, certified nurse midwives, physician

assistants, and mental health professionals, who are eligible for loan repayment funding or have received scholarships, can fulfill their service obligations. The Affordable Care Act included $1.5 billion for NHSC over the next 5 years to expand scholarships and loan repayments for eligible clinicians to work in underserved areas. Some of this new funding can be used to expand access to health care and attract providers to the Gulf Coast region.

The Centers for Medicare & Medicaid Services (CMS) will continue to work with states and stakeholders to monitor health needs or gaps in care where either Medicare or Medicaid could be of assistance. To the extent gaps in coverage, care, or enrollment are identified, CMS will work with the Gulf Coast states to develop solutions that will provide assistance for beneficiaries, and tools and resources for the states to assist in this effort. In addition, CMS has begun working and will continue working on the expansion of Medicaid that is a result of the Affordable Care Act. This could provide additional access to people in the Gulf Coast states.

- *Strengthen healthcare systems' ability to identify and treat exposure-related illness.* Training and technical assistance can be provided in detection, surveillance, prevention, and treatment of oil spill-related illnesses (both physical and behavioral health), to health care practitioners practicing in the Gulf Coast area, HRSA-funded FQHCs, and look-alikes. Additional research and practice efforts could provide the evidence base needed to transform mental health care access and delivery. These training and technical assistance activities can be accomplished by collaborating with grantee educational institutions and/or stakeholder associations, such as the National Association of Community Health Centers, Gulf Coast area primary care associations, and state primary care offices.

  In addition, HRSA-funded area health education centers, geriatric education centers, and public health training centers can assist by developing and disseminating curricula and information relating to the surveillance, recognition, and treatment of health problems that may result from the oil spill. Other methods of disseminating curricula and information may include HHS websites, promulgation through the three Veterans Affairs medical centers in the region and their primary care centers, as well as Department of Defense medical treatment facilities.

D. Human Services

In the aftermath of a disaster, most human service needs result from loss of jobs and income; reduced family, public, and nonprofit support; and increased acute stressors. The reduction in resources caused by the Deepwater Horizon oil spill will likely increase demand for human and social services designed to assist low-income populations, and those with limited English proficiency.[56] Job training, child care assistance, early childhood education programs, senior services, and cash assistance programs are all examples of services that may see increased demand.

In addition to programs designed to support those with economic needs, human services also encompass programs designed to address acute stressors that often increase after disasters. These programs address domestic violence, the abuse/neglect of children and elders, the needs of vulnerable youth, and other emergent concerns.

The level and timing of needs tend to vary across the two types, thus increasing the importance of monitoring key indicators of well-being. In the immediate aftermath of the oil spill, there has already been an increase in the need for programs addressing acute stressors such as domestic violence. Research on the economic determinants of domestic violence indicates that couples under severe financial strain have elevated rates of violence. Similarly, expectations are that rates of child and elder abuse will increase in difficult economic times.

Early evidence from the area is consistent with research predictions. As economic uncertainty grew between April and June, calls from the Gulf States to the National Domestic Violence Hotline increased by 13%. In the state of Louisiana alone, domestic violence hotline calls from two local communities increased by 81% (New Orleans) and 116% (Lafayette) respectively over the same period.[57] Exacerbating the problem is the fact that earnings and tax revenues from the oil and fishing industries are significant sources of financial support for domestic violence programs in the Gulf Coast states. In the aftermath of the spill, programs are reporting significant cuts in funding from states, localities, the United Way, and corporate and individual donors.

By contrast, programs focused on economic dislocation or income loss do not typically see a large initial spike. Instead, demand for these services is expected to increase as the long-term consequences of the oil spill are felt and as individuals exhaust their savings, credit, and available support from family and friends. If oil and fishing related jobs remain scarce, or if other industries experience downturns, then the demand for economic assistance will further increase. The resources being provided by BP and also by philanthropic organizations will soften the impact of a prolonged economic decline in the affected areas. Yet, as it is too early to predict how much those resources will provide, there may also be an increased demand for federal and state services.

Recommended Recovery Actions

Identifying specific recovery activities in the human services area is a significant challenge. It is extremely difficult to predict what the needs will be over the coming years. There is uncertainty about how quickly local industries such as fishing, tourism, and oil drilling will recover. The longer it takes industries to recover, the greater the economic and stress-related needs of local populations. Although some of the human service needs associated with the Deepwater Horizon oil spill will be addressed by BP and other responsible parties through legal processes, there may be some unmet needs.

As a result, the human services recovery plan must necessarily be less precise than anyone would like it to be. Nevertheless, it is clear that federal efforts in the area will consist of three components: monitoring, making the best use of existing programs and funding, and providing new programs and new money if it is appropriated.

- *Monitoring*

  Given the uncertainty about needs, it will be critical that key human services indicators are systematically monitored. The surveillance section of this chapter details a plan for monitoring quantitative indicators of well-being for the affected areas. In addition, local needs must be monitored through less formal methods, such as town halls, reading local newspapers, and engaging state and local leaders. Efforts should continue to be sensitive to the cultural diversity of the affected areas and to encourage all residents to share their experiences and perspectives.

Utilizing multiple methods will provide overlapping opportunities to observe the pace and nature of recovery on the Gulf Coast. If there is evidence of persistent joblessness and other economic challenges, particularly after final claims by responsible parties have been paid to individuals and businesses, then additional action will need to be taken to address poverty and economic hardship, as well as domestic violence and other stress-related problems.

- *Making the best use of current programs and resources*

  Local governments lead and implement most human services programs for individuals and communities, with funding and guidance from the federal government. If there are sustained economic challenges in the Gulf Coast region that are not addressed through available means, then HHS will capitalize upon existing partnerships with state, tribal, local, and nongovernmental social service delivery networks to maximize Gulf residents' participation in existing programs.[58] Moreover, HHS will use these partnerships to identify ways to better tailor existing discretionary funding streams to local needs.

  In 2005, ACF issued a program instruction that illustrated opportunities to use existing federal funding in the TANF program to address significant local needs related to Hurricane Katrina.[59] The instruction included using short-term, non-recurring benefits, and implementing new benefits and services that are consistent with current needs of local residents. This effort serves as an excellent model for any future Gulf Coast region needs.

- *New programs and resources*

  Of course, there are limits to what can be achieved by making better use of existing programs and resources. States and tribes will be faced with simultaneous increases in the numbers of residents meeting eligibility criteria for human services programs, and reductions in tax revenues required to pay for services. New targeted programs and funding may be needed to help residents cope with unmet economic needs and stress-induced problems.

  One example of targeted support is to provide funding to states or organizations to create subsidized jobs. This has been done successfully in the American Recovery and Reinvestment Act (ARRA) through the TANF Emergency Fund.[60] A survey of state officials running these programs suggests that, by September 2010, more than 240,000 unemployed individuals will have work in subsidized jobs funded in whole or in part by the TANF Emergency Fund. Nearly half of these jobs are targeted at youth. However, the TANF Emergency Fund has reached its capacity, and cannot fund new projects unless Congress acts upon pending requests for additional funding.

  Another example is the recent announcement of BP support for expanded state and HHS behavioral health programs and activities.[61] BP pledged "$52 million to help address the immense behavioral health (substance abuse and mental health) needs of Gulf Coast area residents and workers struggling to recover their livelihood in the aftermath of the Deepwater Horizon oil spill." Some of these new funds will be used to implement programming designed to reduce rising levels of domestic violence in the affected areas.

Whether or not these funding needs are met by responsible parties, there are many programs that we expect will experience new funding needs. Needs and programs fall into the following categories:

Direct Benefits:

- Temporary Assistance for Needy Families (TANF): Also known as welfare, TANF is a block grant program with funds dispensed by states to needy families in the form of direct payments, support services, education and training, referrals, employment services, job training, subsidized employment, and child care assistance.
- Low-Income Home Energy Assistance Program (LIHEAP): Assists low-income households in meeting their immediate home energy needs. This will be an important resource if unmet needs extend into the 2011 cooling season.
- Medicaid: Medicaid pays health care providers for the health needs of certain low- income individuals and families who fit into an eligibility group that is recognized by federal and state law.
- Children's Health Insurance Program (CHIP): CHIP encourages states to provide health insurance coverage for uninsured children in families whose income is too high to qualify for Medicaid but too low to afford private insurance.
- Supplemental Nutrition Assistance Program (SNAP): Formerly known as "food stamps," SNAP provides low-income households with electronic benefits they can use like cash to purchase food at most grocery stores.[62] SNAP is the cornerstone of the federal food assistance programs, and provides crucial support to needy households and to those making the transition from welfare to work. State agencies administer the program at state and local levels, including determination of eligibility and allotments, and distribution of benefits.
- Unemployment Insurance (UI): Federal-State Unemployment Insurance Program provides unemployment benefits to eligible workers who are unemployed through no fault of their own (as determined under state law), and meet other eligibility requirements of state law. Unemployment insurance payments (benefits) are intended to provide temporary financial assistance to unemployed workers who meet the requirements of state law. Each state administers a separate unemployment insurance program within guidelines established by federal law. Eligibility for unemployment insurance, benefit amounts, and the length of time benefits are available are determined by state law.
- Women, Infants and Children (WIC): WIC provides federal grants to states for supplemental foods; health care referrals; nutrition education for low-income pregnant, breastfeeding, and non-breastfeeding postpartum women; and to infants and children up to age five who are found to be at nutritional risk.
- National School Lunch Program: The National School Lunch Program (NSLP) is a federally assisted meal program operating in public and nonprofit private schools and residential child care institutions. It provides nutritionally balanced, low-cost or free lunches to children each school day.

Services for Children:

- Head Start: Provides grants to local agencies to provide comprehensive child development services to economically disadvantaged children and families, with a special focus on helping preschoolers develop the early reading and math skills they need to be successful in school.
- Child Care and Development Fund (CCDF): Offers subsidized child care services to eligible families through contracts with providers or with vouchers. This program is critical for many families that are engaged in job search and other efforts to stabilize their lives.

Addressing Abuse and Neglect:

- Family Violence Prevention & Services (FVPSA): Funding to the states, territories, and tribes provides core support to more than 1,675 community-based domestic violence programs.
- Child Abuse and Neglect: Supports activities to build networks of community-based, prevention-focused family resource and support programs through the Community- Based Family Resource and Support Program. Activities also support improvement in the systems that handle child abuse and neglect cases, and improvement in the investigation and prosecution of these cases through the Children's Justice Act.

General Social Services Support:

- Social Services Block Grant Program (SSBG): Provides social services funding to states that is used to assist individuals and families in achieving economic self-support or self- sufficiency.

E. Seafood Safety

The capabilities of state and federal laboratories to receive, analyze, confirm, and report the presence of toxic agents in food, as well as clinical and environmental samples, are essential. Laboratories must have the ability to quickly analyze samples and specimens so that the public health, health care, and environmental health systems can prevent, detect, intervene, diagnose, and mitigate the impacts of the disease, agent, product contamination, or condition. Based on the oil's complex composition and necessary testing to date, there will continue to be a prolonged strain on laboratory capacity for longer-term seafood safety monitoring, including surveillance through laboratory results.

There are public health concerns regarding products regulated by FDA, all of which must be inspected and investigated at an accelerated schedule or as a higher priority than other non-statutory workloads as a direct result of a the oil spill. Delays in these inspections negatively affect the availability of safe food, the reopening of waters, public confidence in the seafood supply, and the return of businesses, their workforces, and contributions to the local economy. A prolonged surge of disaster-specific activities correlates with a demand in labor and overtime at both the state and federal level.

Recommended Recovery Actions

Understandably, the economic recovery and restoration of confidence in the safety of seafood will be a long-term effort. The federal government is committed to this effort and has identified several recovery activities such as long-term surveillance and inspection assignments to ensure that seafood harvested from the Gulf waters is safe. These recovery needs require additional funding support:

- Performing seafood processing inspections and analyzing inspection data to identify seafood safety risks. FDA activities would include developing, coordinating, and monitoring targeted inspection assignments, performing compliance support to review findings, and taking regulatory actions to ensure seafood safety, and providing oversight of firm recalls.
- Continuing to collect seafood samples and analyzing them for the presence of oil spill- related contaminants through coordinated and targeted sampling assignments. FDA would also provide surveillance and compliance support to review findings for regulatory actions to ensure the safety of the seafood supply.
- Continuing to perform diagnostic support, sample collection, and laboratory training for seafood sampling. Amounts include salary and benefits of employees deployed with mobile laboratories. FDA would provide diagnostic support and capacity (including training) through mobile laboratories, and conduct seafood testing at state and federal levels. FDA would also support development of rapid analytical methods for detecting oil contamination in seafood. In addition, the mobile operations will coordinate the participation of leadership and staff in decision making and oversight of incident management.
- Acquiring the necessary equipment, instruments and supplies to develop methods to assure seafood safety, and to develop and enhance protocols for reopening waters currently closed for fish and shellfish operations. These acquisitions would bolster FDA laboratory capacity as well as the capacity and capabilities of the Food Emergency Response Network, which is comprised of both state and federal laboratories.

F. Research

Although there have been 38 major spills involving oil supertankers in the past five decades, including widely publicized disasters involving the *Exxon Valdez, Sea Empress, Braer, Prestige* and others, there are surprisingly few studies on the effects of oil spills on human health. Most studies to date have focused on short-term (acute) effects such as skin, eye, nose and throat irritations, and psychological symptoms. These studies used cross-sectional designs without control groups, making it difficult to interpret the findings. These studies are also limited in that they have been restricted to the relatively healthy oil worker population.

With regard to seafood safety, it is known that the crude oil is a mixture of many different chemicals, a number of which are well established as being harmful to people if ingested in contaminated food or water. Polycyclic aromatic hydrocarbons (PAHs) are toxic components of crude oil that are of high concern if ingested, and the FDA has set limits at a level that is protective of public health. Current science indicates a low risk that dispersants will bioconcentrate in seafood, and a non-sensory method to detect their presence in seafood is in the final stages of development. Research is needed to enhance the capabilities of detecting PAHs and other potential contaminants, including dispersants, associated with the Deepwater Horizon oil spill.

A small number of lab-based studies in animals and cell systems, including seafood (mussels) contaminated with oil mixtures and extracts such as PAHs, demonstrate DNA damage, changes in metabolic and DNA repair enzymes, and hormonal disruption.[63,64] Available data from toxicological and biological studies are grossly inadequate to predict potential health consequences from the Gulf Coast oil spill. Most studies involve dosing with single compounds (crude oil or oil extract) as opposed to the oil/dispersant mixtures that are present in the Gulf, which may have synergistic effects when combined. This is particularly true for oil contaminants that could be absorbed and may bioaccumulate in aquatic life. Just as important, the potential long-term bioconcentration or bioaccumulation of heavy metals are of concern. Validation of methods to detect heavy metals, sampling strategies, and research of how these metals may accumulate will be conducted.

The capability of state and federal laboratories to provide long-term monitoring of seafood safety depends on their ability to develop and deploy more rapid and portable analytical methods to detect oil contamination in the field. Recent technological advances in molecular analysis of genes, proteins, and other small molecules in biological samples also provide a unique opportunity to develop sensitive and informative biomarkers of exposure and early effects in lab-based and human studies.

Recommenced Recovery Actions

There are significant research efforts that can be undertaken to address community health concerns and fill gaps in knowledge about the health consequences of clean-up activities. The research efforts span study of response workers and volunteer populations, as well as study of the general population and at-risk groups including children and pregnant women. Some of the research has initial financial support from HHS and BP, but few can be fully implemented without additional funding. They include:

- Implementing the GuLF Workers Study (Gulf Long-term Follow-up of Workers Study), a large population (cohort) follow-up study of the physical and psychological effects on clean-up workers and volunteers. GuLF, which has some initial funding support from the NIH, will be a prospective health study that will focus on exposure to oil and dispersant products and potential health consequences, such as respiratory, neurobehavioral, immune disorders, and cancer. The study will also evaluate mental health concerns and other oil spill-related stressors such as job loss, family disruption, and financial uncertainties. Workers with varying levels of exposure to oil and dispersants who were involved in oil burning, skimming and booming, equipment decontamination, wildlife clean-up, and shore-line clean-up will be enrolled to evaluate whether exposure levels and certain types of clean-up activities correlate with adverse health effects. The study is being planned and implemented in coordination with other agencies and local Gulf Coast communities. The IOM will plan and convene periodic meetings to provide scientific and community input on the design, implementation, and progress of the workers' cohort study.
- Testing important toxicological endpoints and biological activities and tissue targets for crude oil, dispersant chemicals, and related mixtures that are appearing in the

Gulf and may be harmful to humans if inhaled, absorbed through the skin, or ingested in contaminated seafood or water. HHS agencies, which have some initial funding support, will conduct studies to include a mixture of literature evaluations, analytical chemistry activities, and biological, toxicological and toxicity pathway screens to inform and extend our understanding of the hazards presented by these complex materials. The findings of these studies will help to identify hazards posed by oil and dispersant chemicals from the Gulf that could be addressed in surveillance or long-term research studies.[65]

- Establishing one or more university-community research consortia in the region to assess health effects of the oil spill on local communities. In these consortia, multi-disciplinary teams of scientists would come together to design and implement a series of interrelated studies related to the health effects of the oil spill. Each consortium would be led by a researcher(s) based in a Gulf Coast state impacted by the spill. Community-based participatory research methods would be used to ensure community involvement in all phases of the program, from planning and implementation, to analysis and communication of findings. The program is envisioned as a trans-NIH effort led by the NIEHS and involving multiple NIH Institutes as research partners. The scientific priorities addressed through the program would not be dictated; consortia partners would identify specific scientific questions and topics related to the effects of the oil spill that they would pursue. Possible topics could include physical, chemical, and psychosocial effects and their interactions; maternal and child health; adolescent, child, and adult behavioral health issues; health disparities; human studies that assess exposure to contaminated air, water, and dietary sources of chemical mixtures; adverse effects on the skin and immune function; and toxicologic studies of environmental samples. Studies could involve collection and analysis of human biospecimens, as appropriate. The IOM will plan and conduct a consensus meeting to develop advice on research priorities for studies involving workers and the affected community.
- Addressing health consequences from chronic exposure to environmental chemicals in individuals with specific vulnerabilities due to age, developmental stage, co-morbidities, and immune compromise such as the frail elderly and chronically ill. These projects could be funded as part of the NIH university-community consortia or via a separate funding mechanism.
- Beginning toxicological and biological research on the dermal and respiratory pathways of exposures from oil mixtures, dispersants, and combustion products. Projects could be conducted as part of the university-community research consortia or via a separate funding announcement. Analysis of effects on skin would build on and expand the findings from the workers' study and toxicological analyses. Studies could be initiated to address specific adverse skin manifestations resulting from clean-up or primary exposure to oil or oil dispersants, and include autoimmune diseases such as scleroderma and lupus that are known to have environmental triggers.
- Conducting research to develop more rapid and portable methods to detect petroleum contamination in seafood and to develop methods to detect other potential contaminants associated with the Gulf Coast oil spill. The FDA has begun this research activity with current resources.

In order to fully assess the long-term implications, we would recommend that a committee of experts be convened to provide periodic independent review of the federal response to the oil spill as it relates to the surveillance and monitoring of acute and long-term physical and behavioral health effects of workers and affected communities. The IOM will provide information and advice to HHS on issues such as feedback on the design and progress of the NIH Gulf Workers Study; research priorities such as investigator-initiated research beyond the Gulf Workers Study, including the assessment of children and pregnant women; and on emerging concerns identified by HHS. IOM conclusions and recommendations on research priorities for studies on the health effects of the oil spill on both workers and the public will inform funding decisions by the Gulf Coast Recovery Council.

V. Conclusion

In order to optimize the conditions for a resilient restoration and lasting recovery from the oil spill, a robust federal, state, and local partnership must incorporate health and human services into the larger matrix of human needs that includes economic and ecological safeguards. The strategic framework for health and human services in this report provides a mechanism for the interrelated elements of surveillance, behavioral health, medical needs, human services, food safety, and research. Each area will inform the others in order to identify and address the evolving health risks brought about by this complex disaster.

The recurrent hurricanes and tropical storms in the region have highlighted the remarkable resilience of the people of the Gulf Coast, but have also resulted in damages to its health and social services infrastructure. A well-coordinated approach, based on careful surveillance and reliable research, is essential to move beyond overlapping cycles of disaster response. In this way, effective recovery promotes health security. Strategic recovery activities—with a focus on prevention—will meet the goal of restoring the health and well-being damaged by the oil spill. They will also provide a solid framework for preparedness, response, and recovery to guard against the health consequences of future disasters.

Economic Recovery

I. Introduction

The Deepwater Horizon oil spill was one of the worst man-made environmental disasters in American history and resulted in severe economic dislocations across the Gulf Coast. It shares features with prior major oil spills like the Exxon *Valdez*, but is singular in its scope, scale, and effect upon the human and environmental health of the region. It poses similar economic challenges as a natural disaster, yet is distinct in both the absence of extensive infrastructure damage, and the breadth of its geographic impact. Simply stated, the Deepwater Horizon oil spill presents a unique set of economic challenges for the Gulf Coast, and the response must be calibrated to address the region's needs. But importantly, as the recovery effort progresses, the Gulf Coast will be presented with a unique set of economic opportunities as available recovery funds are dedicated to local economies and provide focused economic activity.

This chapter presents the economic concerns expressed by individuals, businesses and communities across the Gulf Coast, and identifies steps to support a resilient economic recovery. Some of these challenges are immediate and urgent, such as provision of direct compensation and income support. Others are linked to longer-term forces that were present before the spill, including chronic under employment and wage stagnation in the region. Addressing the region's overreliance on certain natural resources, the need to diversify the economy, and the need to build support for clean energy, will require the involvement of numerous stakeholders and persist long into the future.

Plans to restore the Gulf must come from the people of the Gulf. It is absolutely critical that the economic response reflects local priorities, incorporates local input, and builds from local visions for economic restoration. Indeed, the spill occurred at a moment when communities across the region were actively engaged in planning new economic development. Those visions and plans cannot be lost – they must form the basis for all federal, state, local, and private economic development.

II. Background Information

In considering economic recovery needs on the Gulf Coast following the Deepwater Horizon oil spill, it is important to recognize that there is no uniform economic impact of the spill; communities across the Gulf will continue to feel its impact in different ways. These communities face a diversity of challenges, and will require distinct but overlapping solutions. For example, more heavily oiled areas will need to maintain focus on the speed of the clean-up process to ensure the safety of beaches and water. In areas facing less severe physical impacts, challenges may have more to do with protecting the community's image, so that beaches do not become stigmatized in the public eye.

A. PreExisting Economic Challenges in the Region

Several pre-existing challenges will affect the economic recovery efforts in the region, including:

- *Ecosystem conditions:* The economy and the environment of the Gulf Coast are inextricably linked. A variety of ecosystem services contribute to the region's economy. The Gulf of Mexico is one of America's most important fisheries, it contributes to water and air quality, its barrier islands protect against the harmful effects of storms, and it is a recreational destination for millions of Americans. As described previously in the Ecosystems chapter, the Gulf's environment faces numerous underlying challenges that need to be factored into the response to this crisis.
- *Damages resulting from Hurricanes:* Communities across the Gulf routinely suffer the effects of multiple storms with significant economic impact. The residual damage from hurricanes Katrina, Rita, Gustav, Ike, Dennis, Charlie, and Ivan continue to affect Gulf communities. While many on the Gulf Coast consider dealing with

hurricanes part of life, the most recent storms have been particularly severe, degrading the region's economic resilience, and impacting the ability of individuals to rebound due to these prior strains on their resources.

- *Recession impacts:* While hurricane damage and recovery has framed the recent experience of many Louisiana and Mississippi communities, other areas affected by the spill suffered disproportionately in the recent recession and financial crisis. One Florida stakeholder referred to this as a "perfect storm" of bad economic news, with "the closing of businesses, the oil spill, housing, infrastructure, and the economy."

B. Baseline

In 2008, the economies of the five Gulf States supported more than 19 million jobs and accounted for nearly $2.5 trillion, or about 17 percent, of the United States' GDP.[66] Coastal counties in the Gulf States represented approximately 6.9 million total jobs or 36 percent of the region's total employment.[67] Evidence shows that jobs are more concentrated along the coast in some states than in others. For example, nearly 80 percent of the jobs reported in Louisiana are located along the coast, while the portion is less than 20 percent in Mississippi and Alabama. In Texas and Florida, the coast is home to about 35 percent of total state employment. Early indications suggest that among all these jobs, those in travel and tourism, commercial fishing, and oil and gas drilling are particularly affected by the spill.

Travel and Tourism

According to the United States Travel Association, domestic and international travelers to the five Gulf States spent about $145 billion in 2008, and approximately 1.7 million people were employed in travel and tourism.[68] Roughly a third, or approximately 587,000 people, were employed in the coastal counties of the five Gulf States representing almost 9 percent of total employment in these areas.[69] Texas, with 40 percent, and Florida, with 36 percent, accounted for over three quarters of these 587,000 workers. Louisiana, with 18 percent, accounted for most of the rest.[70] In 2008, annual wages for these 587,000 workers amounted to $17.6 billion.[71] Based on the number of people employed in the travel and tourism industry, approximately one third of total tourism dollars, or about $51 billion, were spent in the coastal counties.

Many of those who work in the tourism industry, however, are either underemployed, work for hourly wages, or both. This portion of the population suffers greatly when the industry is impacted. Many small businesses rely upon tourists to survive. Small businesses—those that employ fewer than 500 people—made up all but 15,000 of the over 500,000 businesses in those states. These businesses have been strained by the hurricanes of 2004 and 2005, as well as the economic downturn, which has tapped out their reserves and made them less able to withstand this most recent crisis.

Commercial & Recreational Fishing

In 2008, the dockside value of commercial fishing landings in the Gulf of Mexico was $659 million. More than two-fifths, or $273 million, were recorded in Louisiana and another fifth, or $122 million, was recorded in West Florida. In 2008, between 7,100 and 17,000 people worked as part or full-time workers in commercial fishing or aquaculture. Furthermore, in 2008, total trip expenditures for recreational fishing in the Gulf were nearly $1.5 billion. There were 3.2 million recreational fishermen who took a fishing trip in the Gulf of Mexico region. An additional 8,700 workers were employed in commercial seafood-product preparation and packaging. Taken together, as many as 25,700 people were directly employed in commercial fishing in the Gulf.[72] Annual average wages across the Gulf in 2008 were approximately $23,500 for harvesting, $28,700 for aquaculture and $26,000 for seafood product preparation and packaging.[73]

Oil and Gas Drilling

In 2008, the oil and gas drilling industry employed approximately 180,000 workers in the five Gulf Coast states.[74] The vast majority of these workers were employed with firms located in Texas and Louisiana. More specifically, 74 percent were located in Texas, 23 percent in Louisiana, and less than 2 percent in each of Mississippi, Alabama and Florida. Less than half of the total workers, or about 75,000 workers, were employed by firms located in the coastal counties of the Gulf Coast. According to the Department of the Interior, there are about 80,000 offshore oil and gas production, construction, and drilling workers in the Gulf of Mexico, less than 10,000 of who work on drilling rigs.[75] Average annual wages were approximately $89,000 for drilling oil and gas wells and $77,000 for support activities supplemental to oil and gas operations.

Summary

In summary, as of 2008, along the Gulf Coast there were approximately 587,000 jobs in travel and tourism, between 16,000 and 26,000 in commercial fishing, and 75,000 in oil and gas drilling. Combined, these three industries account for approximately 10 percent of total private employment along the Gulf Coast. The following chart, which summarizes the data presented above, shows the employment in these three key industries by state.

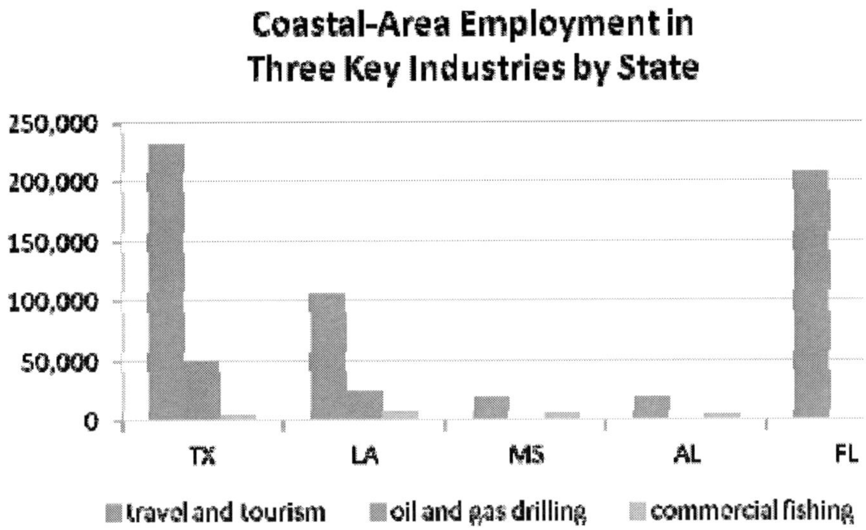

Because so many workers depend on these three key industries, the livelihood of a significant number of Gulf residents has potentially been impacted by the fallout from the oil spill; and because so much of the consumer base in the region is made up of these workers, there have been second and third order impacts on other industries. However, the infusion of funds going toward recovery operations may mitigate some of these impacts as well.

C. How the Deepwater Horizon Oil Spill Further Impacts the Region

Two distinct analytical approaches have helped shed light on the impacts of the spill: an analysis of claims data, and qualitative analysis gathered from numerous visits to Gulf communities have each contributed to a basic understanding of the problems and needs of local Gulf economies.

Claims Data

Until claims processing was transferred to the independent Gulf Coast Claims Facility (GCCF) on August 23, 2010, BP administered its claims process. As they did so, they collected information about claimants, and the justifications that they provided to explain individual claims. The data collected provides insight into the efficacy of the claims process, as well as into the geographic and industrial makeup of claimants. Importantly, the focus here is on claims *filed*, not claims resolved.

Approximately 250,000 justifications were provided by claimants in support of their claims. Given that a claimant could seek compensation for more than one reason, there is not necessarily a one-to-one ratio between claims and justifications offered. Each justification is illustrative of different sources of damage.[76]

As of, August 20, 2010, the last business day before the GCCF took over the claims process, over 150,000 claims had been filed. Individuals and businesses in Florida filed the largest share of claims (26 percent), followed closely by Louisiana (23 percent) and Alabama (20 percent).

Within the four affected states, data on claims shows a concentration of claims in a small number of geographic areas (see Appendix Table X). Specifically, 14 counties accounted for over 60 percent of total justifications. This concentration of activity provides some indication of where the perceived economic impact of the spill has been particularly severe.

Also, the claims data provides some indication of the primary reasons why individuals and businesses lost income or commercial profit. In particular, the data suggest a disproportionate impact on the seafood, tourism and hospitality industries (see Appendix Table X).

Individual and Business Claims Filed				
State	**Total Claims Filed**	**Share of Claims Filed**	**Total Claim Factors**	**Share of Claim Factors Filed**
LA	35,053	23%	72,411	29%
AL	31,210	20%	55,063	22%
FL	39,366	26%	54,883	22%
MS	26,159	17%	38,947	16%
Other	17,119	11%	21,903	9%
TX	5,067	3%	7,596	3%
TOTAL	153,974	100%	250,803	100%

Past Oil Spills Impact on Local Economies

Deepwater Horizon was more than an order of magnitude greater than any other oil spill the nation has faced to date. Because of this, data regarding past oil spills is unlikely to be useful in analyzing the expected extent or duration of its impact on local economies. Other factors, such as the location of the spill, also make the Deepwater Horizon oil spill unique.

However, looking to past oil spills, particularly ones occurring in or near coastal areas, may shed some light on the types of industries that are typically impacted and some of the other factors attendant to oil spills. This section provides a brief summary of some of the research conducted on a handful of coastal oil spills including the Union Oil Platform A, the Amoco *Cadiz*, the Pemex *Ixtoc I*, the Exxon *Valdez,* and the *Prestige* (research and conclusions from sources outside of the United States have not been independently verified).

Historically, tourism and commercial fishing have experienced substantial negative direct economic impacts from coastal oil spills.

In most cases, tourism spending in areas affected by coastal spills declines sharply. It is the duration of the decline that varies by spill. According to one study, normal levels of tourism in Alaska after the *Valdez* spill recovered the following year and "no permanent damage to the tourism industry appears to have occurred."[77] In the Amoco *Cadiz* case, by autumn of the following year there was almost no trace of oil on the local beaches and tourists appeared to return as usual.[78]

Similarly, commercial fishing in the wake of coastal oil spills is typically brought to a standstill in affected areas as fisheries are temporarily closed. Again, the duration of the impact varies. In the case of the Amoco *Cadiz*, most open-sea fishermen were idle for less than two months and commercial fishing levels in 1978, the year of the spill, did not appear to be significantly lower in affected areas.[79] Data on the total fresh fish catch from the Santa Barbara area showed no discernable decrease in 1969.[80] In the case of the *Prestige* oil spill, while landings decreased dramatically in the year of the spill, in each of the following two years overall landings appeared to return to normal levels.[81] In the case of the Exxon *Valdez*, one study suggests the majority of the effects on commercial fishing disappeared after the 1990 season.[82]

Past spills also demonstrate that duration of impacts on commercial fishing can vary by species. For example, aquaculture oysters took longer than a year to fully recover in the case of the *Cadiz* spill. However, in the case of Exxon *Valdez*, effects lasted for over a decade. In the aftermath of the Exxon *Valdez* spill, the recreational fishing and tourism industries bounced back relatively quickly, but the Prince William Sound herring fishery has been closed for 15 of the 21 years since the spill.[83]

Due to many unique factors, the Deepwater Horizon oil spill is expected to present different patterns of recovery than in prior spills. Review of the impact of past oil spills therefore is of limited value. Instead, it is important assessing the potential impacts on the region to look closely at what is happening in specific communities in the region. And whatever the ultimate economic impacts, a critical step in minimizing those effects is to continue efforts and plans to mitigate damage to the tourism and commercial fishing industries.

Community Economic Assessments

First-hand experiences from affected communities can provide some of the most direct insights into the situation on the ground. Insight gained from first-hand experiences and observations came from meetings and conversations with fishermen, health officials, environmental workers, industry leaders, scientists, members of the business community, local leaders, elected officials, and thousands of Gulf Coast residents.

Agencies across the federal government have also reached out to local communities. In particular, an interagency Economic Solutions Team comprised of the Departments of

Commerce, Labor, Homeland Security, Agriculture, the Small Business Administration, and others undertook significant efforts to seek out and listen to local concerns. This included sending 21 teams, comprised of economic recovery specialists from within and without the federal government, to parishes and counties across the Gulf to help guide economic recovery. These and other conversations with local communities helped to identify key needs and concerns, as well as identify a vision for economic recovery. The results of these assessment teams will be released as they are completed.

III. Principles for Economic Recovery

With the goal of helping all Gulf residents, and particularly, those hardest hit by the spill – from low income and traditionally underprivileged communities, to Non-English speaking immigrant populations, to disadvantaged small businesses – economic response and recovery efforts should address these five principles of recovery:

- *Principle 1*: Compensation and income support for affected individuals and businesses are critical to helping communities weather disasters and recover quickly. In the case of the Deepwater Horizon oil spill, direct compensation is in many instances providing the primary form of income replacement. However, the federal government has an important role to play in providing more traditional forms of income support through safety-net programs and by ensuring integrated access to assistance so that there is no "wrong door" for those in need.
- *Principle 2*: Affected communities can benefit greatly from economic assessment, technical assistance, and capacity building. Specifically, analytical capacity and assistance with navigating federal and state programs and non-governmental resources can help recovering communities.
- *Principle 3*: Addressing public perception concerns and mitigating long-term reputational damage in the wake of this spill will be central to economic restoration.
- *Principle 4*: The pace of small business recovery will be central to the recovery of communities across the Gulf Coast. Support for small businesses is a vital component of recovery. Small businesses are critical to the long-term health and economic recovery of the Gulf Coast. As outlined below, small business is one of the drivers of local economies and one of the first catalysts of economic recovery. Small businesses tend to be more vulnerable to shocks than big firms but, with the right tools, small businesses can adapt and benefit from recovery and restoration, too.
- *Principle 5*: Recovering regions should use the opportunities presented by the recovery effort to address longstanding economic and environmental challenges.

The following section describes recent strategies and ongoing activities that fall within these core areas, and outlines recommendations for moving forward.

IV. Recovery Strategy

A. Compensation and Income Support for Affected Individuals and Businesses

Direct compensation and income support obtained through the claims process are critical short-term responses necessary for economic health in the wake of the oil spill. These tools help bridge lost income and avoid further harm to state and local economies.

Yet, across the Gulf Coast, communities have expressed significant concern about the efficiency of the claims system and about their own capacities to use the process. Individual and small business claimants reported that they lacked certain information (e.g., business receipts, sales documentation, etc.) to help them present their claims. Claimants conveyed their belief that these issues complicated the claims process. In response, some communities have requested technical assistance to ensure that individuals and small businesses are able to file claims.

From the earliest days of the spill, the federal government has made it a top priority to ensure that affected individuals and small businesses have "one stop" access to the support services they need. This support has taken the form of assistance with claims or applications for government programs such as unemployment insurance or the Supplemental Nutrition Assistance Program (SNAP or food stamps).

The Integrated Services Team

Early in the response efforts, the National Incident Command (NIC) established the Integrated Services Team (IST), devoted to developing and overseeing a unified approach for the coordination of support to individuals and small businesses impacted by the oil spill. During the early weeks and months of the response, this effort included several key components, including:

- *Working with BP to ensure that claims facilities also provided information about federal benefit programs:* The government presence at claims facilities included federal and state personnel, as well as access to program applications. As a matter of course, claimants could access materials regarding benefit programs, including unemployment insurance, Supplemental Nutrition Assistance Program (SNAP), and Medicaid. The Small Business Administration also had an active presence at claims facilities, offering information about disaster loan programs for qualifying businesses affected by the spill.
- *Facilitating multi-lingual access to assistance:* Early on in the response, the IST recognized that non-native speakers were having difficulty obtaining support. In particular, the Vietnamese community was struggling to access information and gain assistance. The team immediately developed a sub-group to focus on issues of language access. Multi-lingual materials became commonplace at claims facilities and the newly established GCCF has continued that same commitment to language access.
- *Reforming the federal government's one-stop web portal for disaster victims:* As part of the administration's commitment to streamlining access to assistance, the IST worked to create a digital "one-stop" presence, in addition to a network of in-person facilities and phone hotlines. Shortly after the spill began, the IST worked with the

Department of Homeland Security to enhance www.disaster assistance.gov, the federal government's digital platform for disaster victims, which provides information about how to access relevant federal programs. The portal now consolidates access to information on the claims process with access to information on federal and state resources. This new content is available in English, Spanish, and Vietnamese.

- *Easing the transition to the independent Gulf Coast Claims Facility:* Members of the IST served as the primary federal liaisons during the transition to the GCCF—sharing lessons learned from months of experience on the ground.

The Gulf Coast Claims Facility

While the Integrated Services Team initially played a crucial role in coordinating with responsible parties, its role evolved with the development of the independent claims process. On June 16, 2010, the president announced a commitment made by BP to contribute $20 billion into a trust to cover spill-related costs, including compensation and damages resulting from lost economic activity as a result of the spill. BP also agreed to turn over control of the claims process to Kenneth Feinberg, who established the Gulf Coast Claims Facility (GCCF) on August 23, 2010.

The GCCF is an independent claims facility responsible for the collection and resolution of individual and business claims. Individuals and businesses may submit a claim to the GCCF for clean-up and removal costs, damage to real or personal property, lost earnings or profits, loss of subsistence use of natural resources, or physical injury or death.

Recommended Recovery Actions

- *Continue efforts to integrate services and facilitate access to government programs:* The integration of services for disaster victims should and will remain a priority for the administration. While the GCCF is separate and independently run, the federal government will play an ongoing role in ensuring that, through that process, claimants are aware of and have access to federal programs. As individuals, families, small businesses, and communities progress in their economic recovery, a different and more diverse suite of government program materials should be distributed at GCCF facilities and integrated into the www.disasterassistance.gov platform. This would begin with the distribution of materials compiled as appendices to this report. In addition, under this recommendation federal agencies would continue to provide relevant program materials and related information in appropriate languages.
- *Offer resources for tax guidance to those affected by the spill and/or receiving compensation from the GCCF:* The IRS has taken a number of steps to assist taxpayers impacted by the oil spill. To help with questions and issues, the IRS has posted the answers to frequently asked questions on its website and has established a dedicated phone line to deal with taxation questions arising from the oil spill. In addition, the IRS has announced a number of options available to assist taxpayers experiencing hardship due to the spill. Options include postponement of collection actions or added flexibility to coordinate installment agreements. The IRS has also worked with the Gulf Coast Claims Facility to establish voluntary procedures by which taxes could be withheld, as appropriate. Going forward, the IRS is planning

additional outreach to taxpayers during the 2011 tax filing season, which will present particularly complex and challenging issues for many Gulf residents and small businesses. Every effort should be made to provide appropriate educational materials and access to technical assistance, to ensure that taxpayers know of all their options if they are facing financial hardship.

- *Address the financial education and financial access needs of affected communities:* Households affected by the spill might require assistance obtaining financial education resources and accessing financial services that meet their needs, including managing any compensation payments. In addition, small business owners affected by the spill might need specialized credit and counseling support to help them adapt to the changed business environment. Treasury's Community Development Financial Institutions Fund (CDFI Fund) has long supported community development financial institutions (CDFIs) in the region. Treasury-certified CDFIs in the region are already assisting those affected by the spill. Going forward, the CDFI Fund should conduct outreach to local CDFIs to help them expand their impact by increasing awareness of their services. Members of the 21-member Financial Literacy and Education Commission (FLEC), some of which are already working to address financial education needs in the region, should continue to coordinate their efforts so that affected households have streamlined access to available federal financial education resources, including the FLEC's national financial education hotline and website. Finally, Treasury, the CDFI Fund, and the FLEC members should continue the dialog they have already begun with local stakeholders, such as nonprofits, financial counselors, and CDFIs, to explore further opportunities to jointly address community needs. Financial education services should be coordinated with emergency, behavioral, and physical health services, such as such as emergency food assistance.
- *Provide funding to states or organizations to create subsidized jobs:* If Congress acts and provides new dedicated resources, Gulf Coast states could establish subsidized employment programs that would promote job creation and help those displaced by the spill attain new job skills. Four of the five Gulf states – Alabama, Mississippi, Texas, and Florida – implemented subsidized employment programs using TANF Emergency Funds over the past year to provide out-of-work parents (and, in some cases, disadvantaged youth) with the income, dignity, and training that comes with a job.[84] While the new funding could be modeled broadly on the successful TANF Emergency Fund, some changes would be needed to reflect the unique situation of the Gulf Coast. In particular, while subsidized jobs funded though the Emergency Fund were limited to low income parents and youth, a subsidized employment effort to address joblessness in the Gulf region could be broadened to other individuals as well.
- *Increase federal efforts to drive local hiring in response and recovery work:* Already, response efforts have been important drivers of employment in the Gulf. For example, through the Vessels of Opportunity program (VOO), the federal government worked closely with BP to ensure that priority was placed on local commercial assets, and their owners who make their living from the Gulf of Mexico in finding vessels to join the response. VOO employed over 3,200 vessels at its peak.

Economic Assessment and Evaluation teams heard from communities that they perceived VOO to have generated substantial local employment.

The federal government has also encouraged BP to hire locally. To that end, Secretary of Labor Hilda Solis wrote a letter to BP in which she called on the company to have its contractors and subcontractors post all job orders with the states' public workforce systems through each state's One-Stop Centers. Preliminary data indicate that the majority of response contractors – whether hired directly by BP or through federal agencies – are either working within their home states or come from other states affected by the oil spill.

While federal law generally does not permit the federal government to favor organizations, firms, or individuals in particular geographic areas when awarding federal contracts, there is an exception where another statute authorizes or requires that an acquisition be made from a specified source.

To that end, Congress should pass legislation that would allow greater flexibility to prioritize organizations, firms, and individuals residing in areas affected by the spill for federal contracts directly related to the response and recovery efforts. This provision would be modeled after similar provisions in the Stafford Disaster Relief and Emergency Assistance Act, which allows for the prioritizing of local hiring in the event of a presidentially declared disaster. Legislation should specify that such prioritization be permitted for clean-up activities and for longer-term restoration and recovery work—so that these ongoing efforts continue to drive employment and economic prosperity in the Gulf. This would, when combined with efforts to help local small businesses procure contracting opportunities (discussed later), maximize the extent to which the recovery is driven locally.

B. Economic Assessment, Technical Assistance and Capacity Building for Affected Communities

Communities impacted by disasters often benefit from organized community recovery planning. To develop these plans effectively, communities must immediately begin to take stock of their situations and seek to understand both the scope of damage and the desired path to recovery. However, communities often lack the capacity to do this effectively – particularly in the wake of a disaster, when already-limited resources and human capital are often stretched thin. The administrative capacity of local governments is frequently challenged by the additional burdens associated with economic dislocation and recovery.

Meetings with communities and local governments point to multiple challenges associated with insufficient local capacity. For example, there are serious concerns from local government organizations, including special districts, about loss of sales and tax revenues as a result of the spill. Those impacts will be felt for months as receipts come in at rates lower than projected. In addition, some communities expressed a lack of capability to undertake comprehensive data collection efforts that are needed to assess their economic situations. Local governments have also asked for assistance navigating existing federal programs—noting, in particular, that they would benefit from assistance with grant application processes.

The federal government can play an important role in helping to mitigate local challenges and provide additional capacity to disaster-impacted communities, as well as to state and local governments. Federal efforts can support assessment, analysis, and technical assistance in planning and executing economic recovery efforts. Since the spill, the federal government has done this through the deployment of 21 teams of economic recovery specialists to communities across the Gulf. Each team deployed with the mission of helping local communities orient to their situation and develop steps toward economic recovery.

The deployment of these teams was an initiative of the National Incident Command's Economic Solutions Team, in partnership with the International Economic Development Council. The teams included federal agency representatives who, working in cooperation with experienced economic development and disaster recovery specialists, provided vital capacity and technical assistance to local communities. A targeted location for each team's deployment was developed through Geographic Information System (GIS) spatial analysis, and took into consideration demographic, socio-economic, Shoreline Cleanup Assessment Technique (SCAT), and current BP claims data. Ultimately, communities in all five Gulf States will benefit from these assessments.

Before the teams arrived on site, they engaged with community stakeholders to define the problems faced by each community. When deployed, the teams worked with local leadership from state governments, local schools, workforce development organizations, chambers of commerce, economic development organizations, and business representatives from key industries. The teams discussed issues ranging from infrastructure challenges to business recovery needs and concerns about credit and financing. They conducted meetings focused specifically on the status of oil and gas production, commercial and recreational fishing, and tourism. Further discussions were focused on identifying proactive steps to address the specific challenges of individual communities.

Recommended Recovery Actions

- *Build economic development strategies around community needs.* After the initial visits, assessment teams are developing reports to share with the communities. The completion of the reports should and will be intended as further technical assistance in support of communities' individual recovery planning efforts. The reports will stress the importance of empowering communities so that they can take responsibility and guide their own recoveries. At the same time, the reports are expected to encourage communities to coordinate with each other and to develop regional, rather than purely localized solutions.
- *Develop a "toolkit" to help communities understand and access federal programs:* Assessment teams have heard a variety of concerns about access to federal programs, ranging from questions about grant application processes to uncertainty about which federal programs can be utilized. In order to meet these concerns, the federal government has begun developing a recovery "toolkit" that identifies federal programs that may be relevant to communities in different aspects of recovery. The toolkit has been included as an appendix to this report. The information will also be distributed in hard copy across the Gulf, integrated into the www. disasterassistance.gov platform, and publicized through a series of technical assistance workshops and webinars in which federal program representatives will provide local governments and other stakeholders an overview of the contents of the toolkit. Representatives should be available to answer specific questions about different programs and program application processes.

- *Develop Gulf Coast Regional Economic Assets and Economic Clusters:* The president's FY 2011 Budget request supported enhancement for regional economic competitiveness— stating that high-performing regional economies are essential to national growth. The budget supported growth strategies based on stronger regional clusters of economic activity through funding across many agencies. This strategy has significant implications for the Gulf.

  The Gulf Coast is connected to a tremendous range of interrelated economic assets that can serve as the foundation for building economic growth and resiliency. The region should build on its Regional Innovation Clusters (RICs) — geographic concentrations of firms and industries that do business with each other and have common needs for talent, technology, and infrastructure. Several agencies are developing models for using RICs in the region, and their combined efforts will be an important part of the region's economic recovery.

  The Economic Development Administration (EDA) is funding a Regional Innovation Acceleration Network that is expected to develop strategies to complement existing technology-based economic development initiatives. These should build on existing academic and industrial research strengths, and economic priorities in key technology sectors – possibly including coastal restoration technologies, energy production, biomedical technology, and renewable and biomass energies.

  SBA's Regional Clusters Initiative focuses on accelerating small business opportunities in existing regional clusters across the country. SBA is supporting technical assistance, business training, counseling, mentoring and other services that will support job growth and competitiveness of small businesses. SBA's Regional Clusters Initiative harnesses the potential of regional clusters to drive economic growth and strengthen American competitiveness.

  Recently, the SBA announced support for 10 regional economic development and job creation efforts through a new pilot program, "Innovative Economies." This included the Enterprise for Innovative Geospatial Solutions in Mississippi and Louisiana. This effort organizes complimentary geospatial businesses, government agencies, and educational institutions in Mississippi and the Louisiana Gulf Coast to help create a trained workforce, transfer geospatial technology from public institutions into the market, and foster new business growth.

  Going forward, strategies might include: establishing regional innovation organizations to streamline and coordinate technology entrepreneurship, service delivery and management of area incubators, research parks and technology transfer activities, and providing proof of concept funding and early state seed capital to start-up technology firms in the targeted clusters.

C. Rebuilding Public Support for Affected Industries

Both natural and man-made disasters can affect an entire region's image, as well as the image of its states, communities, and core industries. Disasters inevitably dissuade some tourists for some period of time depending on the duration and severity of the crisis, thereby compounding the economic and environmental damage caused by the event itself.

As described earlier in this section, communities in the Gulf are already reporting that perceptions of the Gulf, its beaches, and the seafood industry are having a negative effect on tourism and recreational activities. "We're losing vacationers because of the perception and may lose them permanently," said one stakeholder in Escambia County, Florida. Communities worry both about negative perceptions and simultaneously about real damage being marginalized – "we actually had oil on our beaches, this is not just perception," said another stakeholder in Santa Rosa County, Florida.

While communities remain concerned about seafood and water safety issues, the cleanliness of beaches, and the need for continued testing and monitoring, they are looking for ways to balance these safety concerns with efforts to market core industries and restore the image of the Gulf as a desirable place to live, visit, vacation, and do business.

To date, there have been a number of efforts to bolster the image of the Gulf and its core industries—and some funding has been provided to meet this important goal. At the federal level, efforts have focused on showing support for these industries, and on educating the public about safety information. Senior officials have spoken publicly about seafood safety issues and have addressed public concerns and perceptions.[85] NOAA, EPA, FDA and the Office of the Surgeon General conducted a Seafood Safety Road Show from August 24 through August 28. Commercial fishermen, recreational fishermen, charter boat operators, and local leaders attended stops in Louisiana, Mississippi, Alabama and Florida. To further support the tourism industry, the National Incident Command's Economic Solutions Team worked with states to develop a travel and tourism page on www.restorethegulf.gov to help disseminate appropriate information about open beaches and tourist attractions to the public.[86]

According to its most recent reports, BP has contributed nearly $90 million to support tourism in four Gulf states: Florida, Louisiana, Mississippi, and Alabama – as summarized below:

	BP tourism grant funding by state (millions)
AL	$22
FL	$32
LA	$15
MS	$18

[87]

Most of this funding was provided directly to states, to use or allocate at their discretion—"for the Governors to distribute as they see fit to promote tourism."[88] Florida, for example, has recently announced a "Back to the Beach" tourism advertising campaign.[89]

A minority of the funding was provided directly to communities – including a $7 million grant to seven coastal counties in northwest Florida.[90] Some recipients have used funds to provide financial incentives to tourists. For example, Santa Rosa County, Florida provided partial reimbursement to travelers who rented hotel rooms or condominiums.[91]

This funding – combined with federal efforts described above – was an important step toward repairing the image of affected industries. However, communities expressed concerns to assessment and evaluation teams that it was insufficient, overly weighted toward the interests of states rather than localities, and that their ability to make use of available funds was curtailed by near-term deadlines.[92] Further, the concerns that communities shared with assessment and evaluation teams suggested uneven capacity for developing and implementing strategic programs on short order.[93]

Recommended Recovery Actions

- *Grants for seafood marketing:* The supplemental bill provided $15 million for fishery disaster assistance under the Magnuson Stevens Fishery Management and Conservation Act. NOAA is working closely with the Gulf States Marine Fisheries Commission (GSMFC) – along with their state representatives from Louisiana, Mississippi, Alabama, and Florida – on plans that are intended to help the local seafood industry and the sport fishing community restore national confidence in Gulf fishery products.

  The funding will support three areas to rebuild demand for Gulf products: 1) developing a national Gulf fishery education, outreach, and marketing and advertising campaign; 2) establishing a means to trace fishery products from the point of harvest all the way to the market; and 3) working with conservation organizations on getting sustainability certifications for Gulf harvested fishery products. Together, these three activities should help rebuild confidence in the safety, and increase the demand for Gulf fishery products and sport fishing. This would, in turn, help fishermen, processors, services and marine suppliers, restaurateurs, recreational fishermen, charter boat operators, and sellers of Gulf fish products.
- *Fishery disaster assistance:* The state of fisheries and fishery-dependent businesses should continue to be monitored in order to identify and address potential challenges as they evolve. Fisheries and fishery-dependent businesses with valid claims should continue to present them to the responsible parties. The recently passed supplemental will make it possible to provide additional support to mitigate the economic impact on fishermen and fishery-dependent businesses affected by the Deepwater Horizon oil spill in the event that other funding sources prove insufficient. While the hope is that this funding will not be needed, as it represents an important contingency option, its inclusion in the recent supplemental was still an important step in helping to assure the continuance of a robust Gulf fishing industry.
- *Tourism and marketing should remain priorities if future funds become available:* This report encourages the dedication of a significant amount of any Clean Water Act penalties that are obtained to support the region's recovery. In the event that Congress acts on this recommendation or other funding becomes available, tourism and marketing should be priority investment areas. Moreover, future funds should adhere to lessons learned from assessment and evaluation teams and other reports about the effectiveness of tourism and marketing efforts to date. Specifically, future funds should support efforts that:

  Place greater priority on the needs, interests, and limitations of localities and communities. Funding guidelines should specify the importance of community-based efforts—such that communities receive both funds and guidance about best practices for efficient program implementation. Recipient communities should be encouraged to work together both to share best practices and to undertake efforts that are complementary.

  Establish greater transparency about the availability and use of funds. This should begin with clear and publicly known standards for the distribution of funds.

Recognize that tourism and marketing needs will not end with this summer's tourism season. The funds provided to date by BP came with specified deadlines for use of funds that were established in memoranda of understanding between BP and recipients. These deadlines stipulated that funds had to be used by mid-to-late September. Communities have expressed significant concern about the longer-term fate of their core industries, and funding guidelines should be structured to account for a timeframe that extends beyond this immediate summer season.

D. Targeted Small Business Support

Communities report that small businesses are challenged by the uncertainty of current circumstances. In particular, concerns about the future are contributing to a general wariness about whether or not to take on loans and invest in future business opportunities on the Gulf Coast. Local business leaders have said that small businesses are reluctant to hire additional employees or extend significant resources to acquire inventory for the next season. They are also concerned that economic uncertainty is dissuading investment from outside the region. Finally, local banks report difficulty extending operating credit to businesses for payroll and working expenses.

As one of the drivers of local economies and one of the first catalysts of economic recovery, small businesses are critical to the long-term health and economic recovery of the Gulf Coast. There are 3 million small businesses in the Gulf, making up 97.6 percent of all employers and employing 49.4 percent of all workers. As of 2006, small businesses accounted for 54.1 percent of private sector employment in Louisiana, 50.1 percent in Mississippi, 49.7 percent in Alabama, and 44 percent in Florida. To this end, the Small Business Administration (SBA) has fielded three types of assistance programs to small businesses in the Gulf region.

First, SBA has been offering Economic Injury Disaster Loans (EIDLs) for impacted small businesses in parishes in Louisiana since May 6, and counties in Alabama, Mississippi and Florida since May 14. EIDLs offer up to $2 million to assist with ongoing operating expenses at a 4 percent interest rate with a term of up to 30 years and deferment of the first payment for 11 months. The loans may be used to pay fixed debts, payroll, accounts payable, and other bills that can't be paid because of the oil spill. As of September 15th, 311 EIDLs worth $27.1 million had been provided (Alabama $5.1 million; Florida $7.8 million; Louisiana $12.1 million; Mississippi $2.2 million).

Second, many small businesses and their employees in the Gulf region have existing disaster loans following Hurricane Katrina and the other storms that have impacted the Gulf area over the past few years. As a result, SBA is also offering deferments on these existing disaster loans for eligible small businesses and their employees impacted by the spill. These deferments allow small business owners, as well as their workers with disaster home loans, to keep more cash in their hands while dealing with the financial crisis caused by the oil spill. As of September 15th, $113.9 million in homeowner and business disaster loans, and $5.2 million in monthly payments, had been deferred.

Third, following the spill, the SBA established 31 Business Recovery Centers in Louisiana, Mississippi, Alabama, and Florida staffed by 95 SBA Disaster Assistance employees and 30 Small Business Development Center staff.[94] At the centers, online, and on the phone, staff provides counseling and technical assistance to small businesses on the SBA loan application process, the claims process, and other options open to small businesses for assistance.[95] Additionally, the staff assists with translating SBA and claims forms into a number of different languages, including Spanish, French and Vietnamese. To date, SBA and SBDC staff have completed more than 3,939 field interviews and taken more than 1,935 calls at the Disaster Assistance Call Center.

In addition, USDA has provided capital to small farms, aquaculture, rural infrastructure (Electric, Telecommunications/Broadband, Water and Environmental Programs); as well as a diverse portfolio of small businesses, community facilities, housing and other targeted sectors like bioenergy and renewable energy through its Farm Service Administration and Rural Development Programs.

In response to the sustained damage to the Gulf Coast economy, the administration has already spent nearly $144 million in contracts to respond and recover from the oil spill, coming from 11 federal agencies. Of this, close to $41 million in contracts have gone to small businesses (~29% of contracts), which exceeds the federal small business contracting goal of 23%.[96]

Recommended Recovery Actions

- *On-site small business counseling*: Across the five Gulf States, SBA has established five Small Business Development Centers, 15 Women's Business Centers, and 25 SCORE chapters with a total of more than 950 counselors operating on the ground.[97] Beginning in Q1 FY2011, SBA's partners will conduct in-person, on-site training in many of the affected communities, and assist small businesses with financial matters, business filings, and business planning needs, including new market opportunities. SBA should coordinate access to, and leverage business and trade promotion opportunities being put together by, other federal agencies.
- *Small business federal procurement opportunities*: Beginning in Q1 FY 2011, SBA is planning to work with the General Services Administration (GSA) and other federal agencies to identify federal procurement opportunities for small businesses in the five state region. A special webpage will link small businesses to opportunities in their regions. SBA should work closely with GSA to identify opportunities on the GSA schedule that could be available to small businesses located on the Gulf Coast.

  This will also help facilitate local hiring as discussed earlier. In addition, SBA should work closely with state governments and nonprofit organizations to market these opportunities to small businesses.

  Throughout FY 2011, SBA, GSA, DOD, and DHS are expected to conduct special outreach and training sessions – both in person and via webinar – on how to do business with the federal government, in order to increase the capacity of small businesses seeking federal contracting opportunities. For those small businesses already doing business with the federal government, SBA plans to conduct matchmaking conferences that would bring together federal prime contractors looking to hire subcontractors. These conferences should be held throughout FY 2011.
- *Assuring access to capital:* The Section 108 Loan Guarantee Program is the loan guarantee provision of the Community Development Block Grant (CDBG) program, administered by HUD. Section 108 provides communities with a source of financing

for economic development, housing rehabilitation, public facilities, and large-scale physical development projects. This particularly flexible form of federal funding allows communities to transform a small portion of their CDBG funds into federally guaranteed loans large enough ($5 dollars for every $1 in CDBG funds) to pursue physical and economic revitalization projects that can renew entire neighborhoods. Local governments borrowing funds guaranteed by Section 108 must pledge their current and future CDBG allocations to cover the loan amount as security for the loan.

- Gulf Coast states can make applications to HUD now for Section 108 Loans for small business assistance such as: working capital loans, real estate acquisition/development/expansion, and/or equipment purchase loans. Funding is non-competitive and is available throughout the year subject only to annual congressional approval of loan guarantee authority.

E. LongTerm Economic Diversification/Resilience

While the immediate efforts in response to the spill have centered on compensation, assessment, marketing and other near-term priorities, a central component of the economic vision for the Gulf Coast involves diversifying the economic drivers that create jobs and wealth in the region. This is important, not only to avoid volatility in the event of future disasters, but also to increase job growth and prosperity across the region.

Communities across the Gulf Coast have almost uniformly articulated a strong desire to diversify, oftentimes noting that the spill revealed the fact that their economies are too narrowly concentrated. Many communities had plans to diversify before the spill occurred (and they expressed frustration that recent progress toward realizing projects such as airports and shipping docks might be stymied by the spill). Moreover, communities articulated the need for workforce retraining to support economic diversification.

Disasters can provide opportunities to address long standing challenges and to reconsider fundamental questions about a community's or a region's economy and future direction. By incorporating preexisting ideas about diversification into the long-term economic response, the Gulf can use its valuable assets to move toward a 21st century economy and workforce.

Potential Section 108 Loan Guarantee Authority for Gulf Coast States

NAME	CDBG10	5 X CDBG Allocation
ALABAMA STATE PROGRAM	$ 28,171,248	$ 140,856,240
FLORIDA STATE PROGRAM	$ 29,565,984	$ 147,829,920
LOUISIANA STATE PROGRAM	$ 30,533,587	$ 152,667,935
MISSISSIPPI STATE PROG	$ 32,946,330	$ 164,731,650
TEXAS STATE PROGRAM	$ 79,264,729	$ 396,323,645
Total	**$ 200,481,878**	**$ 1,002,409,390**

The challenge in this area is to balance short-term spill responses with preexisting long-term goals. The vision of a healthy, diverse, 21st century Gulf economy has great potential in several key areas that are vital to the Gulf Coast's long-term success. But, ultimately, the decision of whether and how to diversify an economy must be the decision of, and driven by, the local community.

Recommended Recovery Actions

- *Invest in export potential*: Exports are a key piece of America's economic recovery. This is especially true for the Gulf Coast. In increasing exports, the Gulf Coast can diversify sources of capital and income and reduce reliance on the traditional Gulf industries. Furthermore, diversification of exports means a more diverse and better trained workforce, with more income to bring back to the region.

  In the short-term, programs like the ongoing ExporTech Initiative have been designed to assist participating companies in developing an international growth plan. The initiative provides experts to vet the plans, and connects companies with coaching and access to organizations that can help them move quickly beyond planning to export sales. Based on the program's experience in 19 states, companies in the initiative generally realize sales overseas very rapidly, often within three to six months.

  In the long-term, the Gulf region can play a vital role in the broader National Export Initiative and its goal to double exports in the next five years. Businesses across the region can take advantage of new trade opportunities in a broad range of industries at export.gov.
- *Support development of a 21st century workforce:* A second measure to ensure a diversified gulf economy is job training efforts that will enable workers to enter new and potentially lucrative careers in a clean energy economy. In the short-term, the Department of Labor awarded $27 million in National Emergency Grants (NEGs) to workforce agencies in Alabama, Florida, Louisiana, and Mississippi to fund retraining and employment services. NEGs are used to temporarily expand service capacity at the state and local levels in response to significant dislocation events. They can support workforce redevelopment needs that emerge from a wide range of shocks—helping displaced workers gain new skills so that they can compete in thriving and emerging industries. NEGs can provide the additional resources necessary to assist states and communities with developing retraining opportunities in partnership with institutions like community colleges and technical schools. The Economic Assessment Teams routinely heard from communities about how much they relied on the NEG program to help support job training—and how much they think they could benefit from additional NEG funding and greater flexibility in the program.

  In May 2010, the administration requested supplemental funding to support additional workforce retraining through the National Emergency Grant program. Though the request has not yet been enacted by Congress, the administration supports passage and should work in partnership with affected communities to promote this important measure and ensure its importance is widely recognized. The administration's request also proposed authorization language that would allow for additional program flexibility typically available only in the event of a Stafford Act disaster declaration, but that has not been available for NEGs issued in response to the oil spill. Once again, this request for additional flexibility resonates with what communities want.
- *Promote a clean energy economy:* Residents of the Gulf Coast have expressed a desire to transition to a clean energy economy. Some areas of the Gulf are well

positioned to foster clean energy investments. The Gulf has tremendous wind resources, and pilot projects utilizing the power of the Mississippi River to generate power hold promise.[98] On the other hand, some Gulf Coast states have not positioned themselves well for a transition to clean energy. For instance, Louisiana, Mississippi and Alabama all rank in the bottom 10 among the 50 states in their adoption and implementation of energy efficiency policies.[99] The success of pursuing a clean energy economy will depend, in large part, on the dedication of state and local communities to invest in the necessary infrastructure for the industry to grow and to tailor it to the unique assets of the region.

One area for potential growth is energy efficiency retrofitting, which has the advantage of requiring relatively modest investments in equipment or expertise. A number of efforts from both the public and private sectors have helped prime the pump for additional investments in this area. For instance, the Department of Energy recently awarded the Southeast Energy Efficiency Alliance $20 million to implement energy efficiency programs in 13 southern cities including New Orleans, Louisiana; Huntsville, Alabama; and Jacksonville, Florida. Alabama received $3 million in grants for energy efficiency retrofits in homes, Texas and Mississippi each received $500,000 for state energy efficiency projects and tens of millions of dollars were awarded to Gulf Coast states for industrial energy efficiency projects also by the DOE.

Alabama will also use its $22 million in American Recovery and Reinvestment Act (ARRA) State Energy Program funding to promote energy efficiency of businesses, schools, and correctional facilities, and to develop renewable energy resources in the state. In addition, by the end of September, CenterPoint Energy – headquartered in Houston and serving all Gulf States – will soon hit full speed on the smart meter deployment project, achieving 900,000 units installed by the end of September, with a goal to reach 2.2 million in 2012. It is now ramping up its Intelligent Grid initiative that will upgrade its substations and power lines to sense problems in seconds, dispatch crews with great precision, and re-route power around problem areas. These changes will not just make meters more accurate; they will improve grid performance throughout the Gulf and save 294 jobs. These programs are underway and are scheduled to scale-up over a period of two to three years.

Building on these efforts, the president has called on Congress to pass a new 'Homestar' program of rebates for consumers who undertake energy efficiency renovations in their homes. This $6 billion program has the potential to save consumers in the Gulf and across the country hundreds of dollars per year on their electricity bills, while both putting people back to work in the hard hit construction and manufacturing industries and also reducing energy consumption.

Biofuels and biomaterials production also represents a promising area of growth for the Gulf Coast region. Gulf Coast states have an established biofuel production industry that could be expanded, as well as opportunities to develop advanced biofuel and biomaterials industries. In particular, Gulf Coast states are relatively well positioned to become leaders in algal biofuels, which are considered a promising source of biofuels by science and industry leaders. These fuels require refining processes similar to those used for petroleum after they are produced and harvested.[100] The Gulf Coast region is home to the largest US industry base in petroleum refining (Gulf Coast states represented 48 percent of US refining capacity in 2009),[101] and is therefore in a unique position to leverage its existing infrastructure and expertise to become a leader in the production and processing of algal biofuels.

The federal government and private industry are investing heavily in this area with long- term growth in mind,[102] including the Department of Energy, which has heavily emphasized investments in the development of biofuel production technologies. Some investments in this area that have already reached the Gulf Coast include: a DOE award of $2 million for biomass feedstock cultivation with project sites in Mississippi and Alabama; $4.5 million awarded by the DOE to Auburn University in Alabama for design and demonstration of systems that harvest, handle and transport woody biomass for cellulosic biofuels; an American Recovery and Reinvestment Act award of up to $1.6 million for expanding fueling infrastructure for ethanol blends to Florida and Texas, among other states; and a DOE award of $18 million to Novemer, Inc., for advanced production of materials from CO2 with a site commitment secured in Baton Rouge, Louisiana.[103]

Shipping ports in the Gulf Coast offer an opportunity to introduce and expand clean energy technology as well. Shipping ports are a major source of air pollution and greenhouse gases as a result of their high energy intensity. Implementation of clean energy technologies to upgrade and retrofit conventional port equipment creates jobs and is highly effective at improving economic efficiency and reducing harmful air emissions. Port handling equipment runs mostly on conventional diesel, but can be replaced or retrofitted to operate on biofuels or cleaner burning natural gas. The Environmental Protection Agency (EPA) has made ARRA awards of over $11 million to the ports of Houston and Corpus Christi to improve emissions of both on- and off-road diesel engines operating at the ports.[104] There are also opportunities to add electrical infrastructure at ports that allow for cargo ship hotel services to be powered by shore side electricity (a process known as cold ironing), rather than by on-board diesel engines. Seven of the 10 largest shipping ports (on a tonnage basis) in the US are located in the Gulf Coast region,[105] and this is an area in which state and local governments should consider investing.

The ARRA has provided the Gulf Coast with the initial seed money needed to make sure that clean technology and clean jobs are an integral part of the 21st century Gulf economy. Examples include: the Texas Gulf Coast area has received $386.3 Million for wind projects, including $178 million to Texas Gulf Wind, LLC in Armstrong, Texas.[106] Louisiana Tank, Inc., in Sweet Lake, Louisiana received a clean energy manufacturing tax credit of $5 million for Geothermal Demonstrations to show the feasibility of a geo-pressured power plant in Cameron Parish.[107] Jabil Circuit in St. Petersburg, Florida was awarded a clean energy manufacturing tax credit for $20.4 million to retrofit an existing plant; to offer solar photovoltaic panel assembly, logistics, procurement, and certification services for mono- and multi-crystalline photovoltaic cell manufacturers.[108] Twelve different companies along Florida's Gulf Coast received a combined $410,000 for solar electricity projects.[109] Shaw Modular Solutions, LLC, in Lake Charles, Louisiana received a clean energy manufacturing tax credit of $10.8 million to fabricate modules used in advanced, passively safe nuclear stations. These stations avoid greenhouse gas emissions and the resulting product aids the domestic nuclear energy industry.[110]

To reinforce and grow this foundation in manufacturing, the federal government has also helped create a clean energy market through incentives for deployment of advanced power generation technologies and energy efficiency improvements. ARRA established programs such as the 1603 Grant in Lieu of Tax Credit program; and expanded the renewable electricity production tax credit (PTC) and the federal business energy investment tax credit (ITC) to encourage the deployment of photovoltaics, stationary fuel cells, wind turbines, and microturbine generators. All of these technologies have a US manufacturing base, including manufacturing in the Gulf Coast.

It also accelerated energy efficiency programs such as the State Energy Program and the Energy Efficiency and Conservation Block Grant Program that help individuals and communities lower energy costs, conserve energy and create jobs. The Gulf states could consider implementing or expanding programs such as state Renewable Portfolio Standards, state incentives for deployment of clean energy and energy efficiency, aligning utility incentives with energy efficiency goals, and engaging utilities to institute net metering to compliment federal incentives. This would further establish and grow a clean energy market in the Gulf Coast, and therefore support and expand a manufacturing base for these technologies.

- *If enacted, utilize Gulf Coast restoration funds to support Gulf Coast economic development:* As outlined earlier in this report, we propose that the president ask Congress to designate a portion of any civil CWA penalties collected as a result of the Deepwater Horizon oil spill for Gulf Coast restoration. As part of this overall recommendation, any such funds should focus on areas like those raised in this report that support the region's long-term economic prospects, and that address issues in historically depressed, economically disadvantaged, and distressed communities.

  Specifically, exports, job training, and clean energy should be prioritized in the funding process.

V. Conclusion

The Gulf Coast has weathered a number of recent economic challenges. In many communities across the Gulf, the Deepwater Horizon oil spill exacerbated pre-existing conditions associated with natural disasters, the recession, a fragile ecosystem, and other factors.

No single solution will drive this complex region's economic recovery. Rather, a successful recovery will require balancing immediate, day-to-day needs with long-term, visionary goals; community-driven solutions with the perspective of regional economic drivers and global markets.

In the short-term, this will involve making sure eligible businesses and households are made whole through various compensation efforts. It will mean helping local governments and communities lead the recovery effort by providing coordination and access to the resources the federal government can provide. It also means helping the industries that were the foundation of the gulf economy to rebound through a variety of public and private efforts.

All of this needs to be done with an eye to the future. In the long-term the Gulf is well positioned to take advantage of its location and resources to fuel exports, retrain its workforce, and help lead the way in new clean energy technologies.

NONPROFIT SECTOR RECOVERY

I. Introduction

As described in the previous chapters, the nonprofit sector plays a critical role in the Gulf Coast's recovery from the Deepwater Horizon oil spill. While the capacity of nonprofit organizations[111] on the Gulf Coast has been impacted by the spill, there still exists tremendous potential for nonprofits to affect and influence the recovery effort.

Nonprofits are critical in addressing community needs across the country; they know what is happening at the local level and are able to identify local communities' most pressing concerns, available resources, and potential limitations. On the Gulf Coast, in the wake of storms, the economic recession, and the oil spill, there is evidence of great resiliency in nonprofit operations, but also fragility in meeting future obligations. The success of the restoration effort on the Gulf Coast will depend on the infrastructure of nonprofits and a focus on nonprofit organizations that have demonstrated success.[112]

This chapter details the state of the nonprofit sector prior to the oil spill disaster, the affect that the disaster has had on the nonprofit sector, how nonprofit organizations are responding to local needs, and recommendations for actions and initiatives that will ensure the continued strength of the nonprofit sector and its ability to impact the Gulf communities that are most distressed.

II. Background Information

The nonprofit sector plays a vital role in the recovery of impacted communities and can impact every aspect of ecosystem restoration, economic recovery, and the post-disaster health and human services efforts, particularly of those with the greatest need.

The nonprofit sector is composed of a wide variety of organizations ranging from small, locally based nonprofits to national organizations with extensive experience in disaster recovery and provision of environmental services, community response, and basic human needs. With the help of funders and partners, including the government, nonprofits help jumpstart and sustain recovery efforts and help build enduring capacity by continuing work in an area after other organizations have departed. They play a particularly critical role in the transition between immediate recovery efforts led by federal and state agencies and long-term recovery managed by local governments and communities.

Governments at all levels (federal, state and local) are strong partners of the nonprofit sector, providing approximately 30 percent of the sector's annual revenue.[113] The remainder of nonprofit funding is provided by fees for services and goods from private sources, investment income, private contributions, and other sources.

Nonprofit organizations are critical for ensuring inclusion of all members of the impacted communities, especially those from historically disadvantaged communities; they act as advocates for a wide range of members in the community, such as individuals with disabilities and others who often have heightened access and functional needs such as children, seniors, and other underserved populations. It is crucial that these individuals as well as families receive timely recovery information; participate and understand the recovery process; and have access to adequate resources.

A. PreExisting Nonprofit Challenges in the Region

Nationally, the economic recession is having an impact on the nonprofit sector. According to the Nonprofit Finance Fund only 18 percent of nonprofits expect to have an operating surplus in 2010.[114] Many nonprofits have experienced and are expecting continued cuts in revenue. This economic climate compounds the difficulty of the recovery process for the Gulf region, and introduces significant challenges for nonprofits operations.

Hurricanes Katrina and Rita, combined with the economic recession, have strongly affected the nonprofit infrastructure across the region. In the aftermath of the storms, there was an immediate drop in the number of nonprofit organizations, followed by significant growth in the number of organizations between 2008 and 2010.[115] Accordingly, in the period after Katrina, total revenues for the region's nonprofit sector grew at a faster rate than the nation as a whole. Following national trends, the region's nonprofits experienced a decline in revenue during the recession, albeit less severe than the national average. In addition, most nonprofits operating on the Gulf Coast are relatively small entities with less than $1 million in annual revenue.[116]

B. How the Deepwater Horizon Oil Spill Further Impacts the Region

In the past five years, the nonprofit sector in the Gulf region has increased in size, at a rate equaling or exceeding growth elsewhere in the country. However, this does not necessarily mean that nonprofits on the Gulf Coast now have a greater capacity to serve those communities experiencing increased need. Nonprofits in the Gulf Coast region report a significant resource gap – financial and human – that affects their ability to meet community needs now and into the future.

Nonprofits Challenged by Increased Demand for Services

Based on surveys conducted in Louisiana, Mississippi, and Alabama after the oil spill, between 30-50 percent of Gulf Coast nonprofits surveyed are reporting greater need among clients and a corresponding increase in the amount of services provided.[117] Nonprofits have also observed decreased fundraising, increased stress among residents, and an increased rate of job and income loss among the populations served.

These trends compel nonprofit organizations to provide emergency response and the provision of basic services such as food and unemployment assistance instead of focusing on long-term recovery. While some organizations are thinking about long-term recovery, many nonprofit and community organizations continue to grapple with the day to day needs of increased demands for food, job training, and mental health services– needs that existed prior to the spill, but were also exacerbated by it. As a result, nonprofits have been unable to conduct formal assessments to determine long-term needs and formulate recovery plans. These organizations acknowledged that more information about community needs would be beneficial, but to date they have only been able to focus on providing immediate response services.

Nonprofits Challenged by Limited Philanthropic Support

One of the biggest challenges facing some nonprofits in the Gulf is a lack of monetary donations; adequate resources are not available to cover assistance costs. Unlike other recent natural disasters in the U.S. and abroad, nonprofit and philanthropic leaders are reporting that the oil spill did not generate an outpouring of donations from the American public.

At a conference held August 24, 2010 by the Corporation for National Community Service, nonprofits and other stakeholders gathered in New Orleans to discuss “Transitioning to New Opportunities.” Leaders stated that while some nonprofits were receiving enough donations to support assistance efforts, others were not receiving enough to meet demand.

Nonprofits are Challenged by Shortages of Specialized Staff

Another constraint in meeting the increased demand for services is limited staff, specifically skilled workers required for managing immediate and longer-term recovery. Given the nature of the oil spill’s impact, nonprofits need individuals with specialized skills and expertise – social workers and case managers, financial experts, and claims lawyers.

C. Initial and Current Responses to the Oil Spill

Since the Deepwater Horizon oil spill, the nonprofit sector has responded to environmental, economic, and public health and human needs all along the Gulf Coast.

The Role of Nonprofits in the Environmental, Economic and Health Response and Recovery

As noted in the ecosystem chapter, nonprofits have been working alongside government and private industry partners on restoration and recovery efforts. Nonprofit organizations have provided supplemental services in the form of beach patrols, wildlife rescue and university research teams, and have assisted in supporting the clean-up effort. As the disaster moves into the long-term recovery stage, nonprofits will continue to be community partners in ecosystem restoration efforts.

Community development organizations and social service nonprofits have also been engaged in the economic response. They help displaced workers find new pathways to employment; provide community-level support for emerging industries such as "green" rehab and construction; and support small business development. For example, the Blue Cross Blue Shield Foundation, with support from Louisiana Serve and the Corporation for National and Community Service (CNCS), hosted Coastal Career Fairs in seven coastal parishes in Louisiana from June through late-August.

From a health and human service standpoint, nonprofit organizations have been providing emergency services to those individuals who have fundamental needs for food, clothing and shelter. For example, Catholic Charities, with financial support from BP, has opened five emergency relief centers where fishers and their families can go for food and housing assistance, counseling, and case-management services. So far, the centers have served more than 5,000 people.

The Role of Volunteers and Community Service in the Response and Recovery

More than 5,000 volunteers and community service participants have been involved in the oil spill response and recovery effort. While volunteers have been engaged in a range of activities, the complex nature of oil cleanup, coupled with health and safety concerns, has limited their role. The Deepwater Unified Area Command, under the direction of Coast Guard Admiral Thad Allen, specified early-on that volunteers and unpaid individuals will be restricted from hands-on hazardous waste remediation efforts.

Despite this limitation, local community organizations and state service commissions report receiving an outpouring of volunteers interested in helping the Gulf Coast.

State service commissions across the region have been leading and coordinating the local volunteer response. Service commissions in Florida, Mississippi, Alabama and Louisiana have been actively involved with the Unified Command Centers and have facilitated the flow of information between the government response and community organizations on the ground. Also, they coordinated efforts that extended beyond volunteer engagement and management. For example, the state commission in Alabama is partnering with the Substance Abuse and Mental Health Services Administration (SAMSHA) in directing and managing grants to local behavioral health programs.

The Philanthropic Response

The ability of nonprofits to respond successfully to the urgent needs in the Gulf will not depend simply on their current capacity, or on their partnerships with and support from the federal government and Congress. It will also require continued commitments from the philanthropic sector, which has mobilized and responded to the region's needs. For example:

- The *Louisiana Disaster Recovery Foundation* has awarded emergency grants totaling $117,500 to nine local nonprofit groups that are providing assistance to communities affected by the spill.
- The *Gulf Coast Fund for Community Renewal and Ecological Health*, a project started by Rockefeller Philanthropy Advisors after Hurricane Katrina, has awarded $200,000 in emergency grants to 32 grass-roots environmental groups in the region (as of June 2010). The Gulf Coast Fund is also organizing tours that allow potential donors to see the work and meet the leaders of the groups the fund supports.[118]
- The *Gulf Oil Spill Fund*, created by the Greater New Orleans Foundation on April 30, supports organizations on the ground providing critical services to fishermen and their families. The foundation has been getting daily phone calls and e-mail messages from people around the world who want to help.
- *GlobalGiving* has created a fund on behalf of National Geographic to support the efforts of partner organizations to reduce the impacts of the oil spill. Nonprofit organizations including the Waterkeeper Alliance, National Park Foundation, American Bird Conservancy, and America's Wetland Foundation are mobilizing to reduce and repair damage to fragile coastal habitats caused by the spill.

III. Recovery Strategy

While previous sections of the report highlight the recovery strategies that respond directly to the environmental, economic, and public health and human needs of the Gulf region, this section identifies strategies to ensure a strong and agile nonprofit sector that will continue to be an effective partner in the long-term recovery effort.

A. Enhanced Coordination

Nonprofits, local communities, and the federal government have been in close communication through local incident command centers. The nature of the oil spill forced traditional first- responder organizations, such as the Red Cross and United Way, to collaborate and coordinate across sector silos. For example, environmental health, domestic violence prevention, and financial education programs and organizations are all serving the same constituency.

A comprehensive network of support for vulnerable communities is critically important. Specifically, there are three significant needs:

- *Data and information sharing*: Information on constituent needs and access to services is crucial. As a part of the Hurricane Katrina recovery effort, several databases were established to track individuals receiving services as well as to inventory program and services available to the community. The Coordinated Assistance Network (CAN) is a consortium of disaster response organizations funded by private foundations that provides an online database. CAN results in shared, secure, current information about disaster victims, their needs, and the timely delivery of services – without duplication of effort. For a variety of reasons, this network has not been fully utilized in the oil spill response. A resource like CAN is needed to provide information to government and nonprofit leaders, within and outside traditional disaster response networks, who are seeking to direct resources to those most in need. The response effort requires information sharing that crosses state lines, a need not met by traditional communication networks.
- *Inclusion of state and local entities:* State and local entities do not feel engaged in the process that determines resource allocation and use. Mechanisms and processes for state and local entities to articulate needs, request assistance and participate in the decision making process about funding and recovery strategies and plans need to be in place.
- *Inclusion of community-based and faith-based organizations:* Community-based organizations have a deep understanding of the specific cultural and ethnic communities impacted by the spill; as a result they must be part of the coordinated response. Often these organizations provide a multitude of response efforts including environmental, economic, and public health and human service needs. Coordination efforts should explicitly incorporate these community and faith-based organizations to ensure that programs and services are reaching all who are in need.

B. Increase Public Awareness of the Need

The public can play an important role in addressing the financial resource gap experienced by nonprofit organizations. Anecdotally, nonprofit leaders express concern that the general public does not understand the deep human needs of the region. Public understanding of the region's needs is a precursor to fundraising efforts. While individual organizations play a role to compel prospective donors, coordinating bodies and infrastructure organizations such as state nonprofit associations can also communicate nonprofit needs to the media and public without diminishing the importance for responsible parties to also do their part.

C. Leverage Volunteers and Community Service Assets

The public can address the resource gap through self-organized community service and formal volunteering. As reported previously, volunteers and community service participants are making a difference every day. Neighboring models and skills-based volunteering are powerful solutions that have amplified the human resource capacity of nonprofit organizations. CNCS is providing $1 million in federal grants to the Gulf States to support growth in volunteerism and increase impact. In addition, the agency is committing additional VISTA volunteers (Volunteers in Service to America) to meet the immediate and long-term needs of low income individuals and families in the region. These activities are part of the agency's broader effort to support the long-term recovery by investing in community capacity to identify and respond to pressing issues.

D. Synthesize additional needs assessment

There is a need for comprehensive research about the impact of the Deepwater Horizon oil spill on the region and the opportunities for nonprofits to lead in the recovery. Nonprofit leaders must receive the impact assessments of the economic, environmental, and health and human needs in an easily understood format to effectively inform their actions. Nonprofit participation in the federal and state led recovery effort will be essential. In addition, CNCS should continue to monitor and synthesize impact assessments that are relevant to the nonprofit infrastructure.

E. Enhance fundraising capacity

The biggest challenges in helping nonprofits secure necessary funding are gaining information about funding opportunities and building staff capacity to pursue those opportunities. As part of the broader response and recovery effort, state service commissions have played a coordinating role and have the potential to serve as an information hub and connector between nonprofits, funding sources, government agencies, and capacity building resources such as nonprofit management support organizations. In addition, CNCS is making additional VISTA volunteers available on the Gulf Coast. VISTAs are dedicated to building capacity and often undertake projects that augment the nonprofit sponsor's fundraising capability and capacity. Additional capacity and better communication amongst stakeholders can help nonprofits secure the necessary financial resources to conduct their work.

IV. Conclusion

The nonprofit sector is an important component of the long-term restoration and recovery effort on the Gulf Coast. Since the Deepwater Horizon oil spill, nonprofits have been on the front lines of the environmental, economic, and public health and human needs response. Although the sector is resilient, the recent impact of multiple hurricanes and the economic recession has adversely affected the sector's capacity, potentially limiting its ability to maximize its role in recovery. In supporting local solutions for recovery, it is critical to support the nonprofit infrastructure and direct investment to nonprofit organizations that are effective in meeting community needs.

APPENDIX A. NONPROFIT SECTOR RECOVERY EFFORT

To get a snapshot about the state of nonprofits in the Gulf Coast region that have been affected by these events, this report highlights data from three time periods: July 2005, immediately before Katrina; April 2008, during the post-Katrina recovery period but before the economic downturn; and April 2010, during the economic downturn but before the oil spill.[119] Although the data does not specifically offer a snapshot for pre- and post oil spill, they help us to understand what happens to nonprofits in the Gulf after a devastating event. We focus on data gathered from the Gulf Coast counties affected by the oil spill, including coastal areas.[120] For all three time periods, we also report data specifically for Orleans Parish, Louisiana, where New Orleans is located, as well as data from the South Census region (the South)[121], and data for the nation as a whole.

Nonprofit revenues have been on the decline since the recession.

Since Katrina but before the economic downturn, there was an increase in revenues to nonprofits across the U.S. (25.3 percent), in the Gulf (28.8 percent) and in Orleans Parish, LA (a noteworthy 34.4. percent increase). Nevertheless, since the economic downturn, the entire country experienced a loss of revenue (a 3.5 percent loss for the nation). This loss did not escape the Gulf region, as it also saw a 2.1 percent loss. However, nonprofits in Orleans Parish, Louisiana experienced a staggering 30.0 percent decline in revenue within that two-year period. The data suggests, perhaps surprisingly, that the economic recession has already had a larger-than-expected negative impact on the Gulf region and Orleans Parish area nonprofits that cannot be attributed solely to the devastation of Katrina. Alternatively, the results could indicate that the longer-term effects of Katrina, layered on top of the rising effects of the recession, are having a profound impact on nonprofits in the Gulf, as opposed to just the recession having such an impact.[122] Regardless of the reason for the decline in revenue, responding to increased demand for services due to the oil spill in light of declining revenues is a steep challenge.

Yet, the nonprofit infrastructure shows signs of resilience.

There is some good news, offering some hope for potential recovery of the Gulf Coast non-profit sector. Despite Katrina and the state of the national economy, the number of nonprofit organizations in the Gulf Coast region has grown steadily since July 2005. First, the number of nonprofits in the region grew slowly and then more quickly, from 17,275 organizations in July 2005 to 17,848 organizations in April 2008, and then to 19,809 organizations in April 2010. This rate of growth (14.7 percent between 2005 and 2010) was almost as large as that seen in the South overall (16.8 percent increase) but larger than the national growth rate (13.6 percent). Although Orleans Parish, Louisiana, lost over 350 nonprofits between 2005 and 2008, the sector has grown in numbers since then, and is now less than 100 organizations smaller than it was in July 2005.

In addition, even given the economic downturn, total organizational assets (including, but not limited to cash assets) among Gulf Coast nonprofits have grown substantially since 2005, at an even larger rate than in the South and in the U.S. overall. Asset growth was much larger among Gulf Coast nonprofits between 2005 and 2008 (44.8%) than between 2008 and 2010 (6.7%), but the overall growth rate over this time period was larger for the area (54.4%) than it was for the South (48.7%) and the U.S. (51.9%). The five-year asset growth rate for nonprofits in Orleans Parish was even larger, at 57.9 percent. Statistics on total revenues also suggest that nonprofits on the Gulf Coast have grown at a faster rate (26.0%) than Southern or national nonprofits (22.9% and 20.9%, respectively).

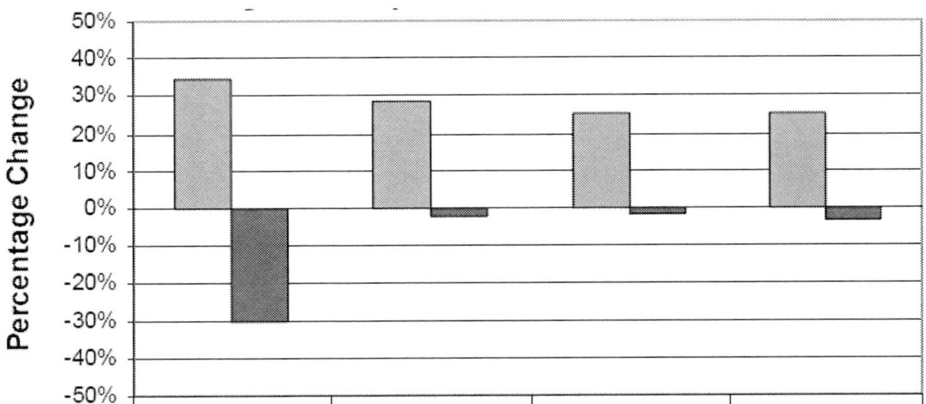

	Orleans Parish, LA	Gulf Coast	South	USA
July 2005 - April 2008	34.4%	28.8%	25.2%	25.3%
April 2008 - April 2010	-30.0%	-2.1%	-1.9%	-3.5%

☐ July 2005 - April 2008 ■ April 2008 - April 2010

Chart 1. Change in Nonprofit Revenues, 2005-2010.

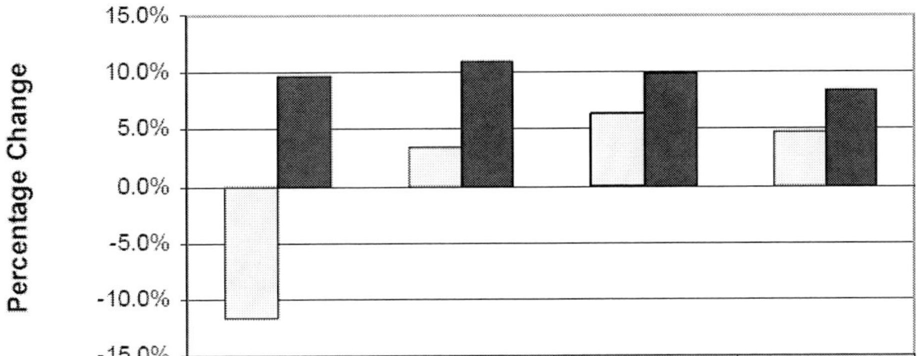

	Orleans Parish, LA	Gulf Coast	South	USA
July 2005 - April 2008	-11.8%	3.3%	6.3%	4.7%
April 2008 - April 2010	9.7%	11.0%	9.9%	8.5%

☐ July 2005 - April 2008 ■ April 2008 - April 2010

Chart 2. Change in Number of Nonprofit Organizations, 2005-2010.

Status of Nonprofits that Address Environmental, Health and Human Needs, and Economic Opportunity

A. Nonprofits Whose Services Will Likely Be Most Needed in the Gulf for Recovery

Data from the BMFs[123] show nationally, nearly 40 percent (over 593,000) nonprofits (out of almost 1.6 million total) are categorized as having a main focus that arguably aligns with the types of services needed in the Gulf after the oil spill[124], as of April 2010. Over 7,700 of these nonprofits are located in the Gulf Coast area and 1,245 are in Orleans Parish, Louisiana. The number of these nonprofits has seen a steady increase in the U.S. and the Gulf since 2005, but a decline and then rebound in Orleans Parish since Katrina. The decline was larger between 2005 and 2008 than it was between 2005 and 2010, signaling that perhaps the devastation of Katrina had a stronger influence on the number of nonprofits than that of the recession.

Assets for nonprofits with these same NTEE codes in the Gulf and U.S. declined after Katrina, but began to climb after the recession. In Orleans Parish, they declined after the economy turned. Revenues, on the other hand, increased between 2005 and 2008 for the Gulf, Orleans Parish and the U.S., but then fell again in 2008.

Table 1a. Mental Health Focus - Number of Nonprofits

				Percent Change		
	Jul 05	Apr 08	Apr 10	Jul 05-Apr 08	Apr 08 - Apr 10	Jul 05 – Apr 10
Orleans Parish, LA	46	36	39	-21.7%	8.3%	-15.2%
Gulf Coast	219	217	231	-0.9%	6.5%	5.5%
South	5,572	5,960	6,429	7.0%	7.9%	15.4%
USA	17,146	18,029	19,012	5.1%	5.5%	10.9%

Table 1b. Mental Health Focus - Total Revenues (in millions)

				Percent Change		
	Jul 05	Apr 08	Apr 10	Jul 05 - Apr 08	Apr 08 - Apr 10	Jul 05 - Apr 10
Orleans Parish, LA	$22.7	$25.8	$27.1	13.8%	5.0%	19.5%
Gulf Coast	$245.6	$292.7	$303.2	19.2%	3.6%	23.4%
South	$5,854.6	$6,598.0	$6,709.6	12.7%	1.7%	14.6%
USA	$24,261.1	$27,936.3	$29,085.3	15.1%	4.1%	19.9%

Table 1c. Mental Health Focus - Total Assets (in millions)

				Percent Change		
	Jul 05	Apr 08	Apr 10	Jul 05 - Apr 08	Apr 08 - Apr 10	Jul 05 - Apr 10
Orleans Parish, LA	$39.4	$41.8	$52.8	6.0%	26.6%	34.1%
Gulf Coast	$167.0	$201.9	$251.2	20.9%	24.4%	50.4%
South	$5,489.5	$6,641.4	$6,620.8	21.0%	-0.3%	20.6%
USA	$17,222.1	$20,996.8	$22,656.0	21.9%	7.9%	31.6%

B. Nonprofits with Mental Health Focus

As discussed in previous sections of this report, given what we have learned from the aftermath of Exxon Valdez and Hurricane Katrina, there may be significant need for mental health services in the region.

According to data from the BMFs, there are over 19,000 Mental Health and Crisis Intervention organization. 230 of these are located in the Gulf Coast area, and 39 in Orleans Parish[125]. The number of mental health nonprofits has grown slightly in the Gulf Coast overall (219 in 2005, 231 in 2010) but has decreased in Orleans Parish, from 46 to 39. Growth rates in both areas lag behind the national and Southern regional rates.

However, while Gulf Coast mental health nonprofits have not grown much in numbers since 2005, they have grown in overall capacity. Assets for Gulf Coast mental health nonprofits increased by 50.4 percent between 2005 and 2010, and revenues increased by 23.6 percent; both rates were much larger than those seen in the South and the U.S.

C. Nonprofits with Environmental Focus

Four hundred and twenty three environmental organizations are located in the Gulf Coast area and 42 in Orleans Parish[126]. The number of environmental nonprofits has grown since 2005, both in Orleans Parish and in the Gulf Coast overall, over the 2005- 2010 time period. In fact, Gulf Coast environmental nonprofits have grown at a faster rate than both national and southern environmental nonprofits, according to all three measures: number of organizations, in total assets, and gross revenues.

Table 2a. Environmental Focus - Number of Nonprofits

				Percent Change		
	Jul 05	Apr 08	Apr 10	Jul 05 - Apr 08	Apr 08 - Apr 10	Jul 05 - Apr 10
Orleans Parish, LA	35	36	42	2.9%	16.7%	20.0%
Gulf Coast	354	378	423	6.8%	11.9%	19.5%
South	14,175	15,243	15,992	7.5%	4.9%	12.8%
USA	31,329	33,655	35,805	7.4%	6.4%	14.3%

Table 2b. Environmental Focus - Total Revenues (in millions)

				Percent Change		
	Jul 05	Apr 08	Apr 10	Jul 05 - Apr 08	Apr 08- Apr 10	Jul 05 - Apr 10
Orleans Parish, LA	$7.7	$9.5	$10.1	23.6%	7.0%	32.3%
GulfCoast	$80.5	$94.7	$86.2	17.8%	-9.0%	7.2%
South	$2,750.5	$3,526.1	$3,627.1	28.2%	2.9%	31.9%
USA	$7,578.8	$10,061.0	$10,319.9	32.8%	2.6%	36.2%

D. Nonprofits with Employment Focus

Four hundred and thirteen organizations with an employment focus are located in the Gulf Coast area, with 71 in Orleans Parish[127]. In the Gulf as well as nationally, the number of employment nonprofits decreased between 2005 and 2008, but has increased since then, though there are still fewer nonprofits in this category in April 2010 than there were in July 2005. Employment nonprofits in the Gulf Coast area saw their assets grow by 56.1% over the 2005-2010 period, which is much faster than the rates seen in the South (30.3%) and the nation (26.0%). However, revenue growth rates for the Gulf and for Orleans Parish lagged behind growth rates for the region and the country for 2005-2010.

Table 2c. Environment Focus - Total Assets (in millions)

				Percent Change		
	Jul 05	Apr 08	Apr 10	Jul 05- Apr 08	Apr 08- Apr 10	Jul 05 - Apr 10
Orleans Parish, LA	$35.6	$45.5	$48.7	27.8%	7.0%	36.8%
Gulf Coast	$135.9	$186.8	$229.6	37.4%	22.9%	68.9%
South	$5,008.0	$6,456.1	$7,250.3	28.9%	12.3%	44.8%
USA	$18,695.6	$25,221.8	$26,694.7	34.9%	5.8%	42.8%

Table 3a. Employment Focus - Number of Nonprofits

				Percent Change		
	Jul 05	Apr 08	Apr 10	Jul 05 - Apr 08	Apr 08 - Apr 10	Jul 05 - Apr 10
Orleans Parish, LA	83	61	71	-26.5%	16.4%	-14.5%
Gulf Coast	419	399	413	-4.8%	3.5%	-1.4%
South	11,029	10,808	10,846	-2.0%	0.4%	-1.7%
USA	39,688	38,620	39,520	-2.7%	2.3%	-0.4%

Table 3b. Employment Focus - Total Revenues (in millions)

				Percent Change		
	Jul 05	Apr 08	Apr 10	Jul 05 - Apr 08	Apr 08 - Apr 10	Jul 05 - Apr 10
Orleans Parish, LA	$45.3	$14.9	$45.8	-67.2%	207.9%	1.1%
Gulf Coast	$171.3	$186.3	$183.5	8.7%	-1.5%	7.1%
South	$8,861.3	$10,440.1	$10,254.8	17.8%	-1.8%	15.7%
USA	$28,977.7	$33,394.5	$32,886.7	15.2%	-1.5%	13.5%

Table 3c. Employment Focus - Total Assets (in millions)

				Percent Change		
	Jul 05	Apr 08	Apr 10	Jul 05 - Apr 08	Apr 08 - Apr 10	Jul 05 - Apr 10
Orleans Parish, LA	$30.1	$14.3	$38.5	-52.5%	169.2%	27.9%
Gulf Coast	$109.7	$138.5	$171.2	26.2%	23.6%	56.1%
South	$9,595.9	$11,927.9	$12,508.0	24.3%	4.9%	30.3%
USA	$30,583.7	$35,945.0	$38,527.4	17.5%	7.2%	26.0%

E. Nonprofits with Financial Counseling Focus

There are a small number of nonprofit organizations whose primary focus is listed as financial counseling – only 772 nationwide, as of April 2010, and only 11 in the Gulf Coast area. One of the three nonprofits in Orleans Parish that specialized in financial counseling apparently ceased operations between April 2008 and April 2010. It was by far the largest of the three organizations with reported assets of almost $3 billion, and appears to have experienced a sharp decrease in its revenues between 2005 and 2008.

Table 4a. Financial Counseling Focus Number of Nonprofits

				Percent Change		
	Jul 05	Apr 08	Apr 10	Jul 05 - Apr 08	Apr 08 - Apr 10	Jul 05 - Apr 10
Orleans Parish, LA	5	3	2	-40.0%	-33.3%	-60.0%
Gulf Coast	12	12	11	0.0%	-8.3%	-8.3%
South	283	290	314	2.5%	8.3%	11.0%
USA	708	755	772	6.6%	2.3%	9.0%

Table 4b. Financial Counseling Focus - Total Revenues Number of Nonprofits

				Percent Change		
	Jul 05	Apr 08	Apr 10	Jul 05 - Apr 08	Apr 08 - Apr 10	Jul 05 - Apr 10
Orleans Parish, LA	$2.4	$1.4	$0.03	-43.3%	-97.6%	-98.7%
Gulf Coast	$3.4	$2.2	$0.79	-37.0%	-63.3%	-76.9%
South	$431.3	$269.4	$334.9	-37.5%	24.3%	-22.4%
USA	$857.6	$913.3	$732.4	6.5%	-19.8%	-14.6%

Table 4c. Financial Counseling Focus - Total Assets Number of Nonprofits

				Percent Change		
	Jul 05	Apr 08	Apr 10	Jul 05 - Apr 08	Apr 08 - Apr 10	Jul 05 - Apr 10
Orleans Parish, LA	$3.31	$3.10	$0.14	-6.3%	-95.4%	-95.7%
Gulf Coast	$4.38	$4.02	$1.19	-8.2%	-70.3%	-72.7%
South	$387.38	$341.30	$430.19	-11.9%	26.0%	11.1%
USA	$2,322.83	$2,120.36	$1,824.08	-8.7%	-14.0%	-21.5%

It is unclear how much the absence of this organization may have reduced the capacity of nonprofits in the area to provide financial counseling services.

F. Nonprofits with Community Improvement and Capacity-Building Focus

Nonprofit organizations whose primary focus is listed as Community Improvement and Capacity-Building are essential to communities as they develop and implement local, cross-cutting self-help strategies for economic and community recovery. There are approximately 143,000 nonprofits in this category nationwide, as of April 2010, and only 1,928 in the Gulf

Coast area. The number of nonprofits with these focus areas have remained relatively unchanged between 2005 and 2008, but showed an increase between 2008 and 2010. Orleans Parish is the exception, with the number of nonprofits decreasing after Katrina, but not since the recession.

Assets for nonprofits in Orleans Parish, the Gulf and the U.S. have increased noticeably since 2005, although the nationwide increase (76% between 2005 and 2010) outpaced that of the Gulf (41.8%) and Orleans Parish (34.7%). Revenues show a similar trend, in that there were notable increases between 2005 and 2010, except for the U.S., which remained relatively unchanged between 2008 and 2010. The Gulf Coast and Orleans Parish's growth in nonprofit revenue outpaced that of the U.S. between 2005 and 2010 and even between 2008 and 2010; in fact, the revenues for community improvement nonprofits nationwide had a negative percent change between 2008 and 2010. These factors together perhaps demonstrate the particular importance of nonprofits that focus on community development and capacity-building and the role they can play in the recovery effort. Their increased revenue in the Gulf could have been a result of donations and grants given to those types of organizations as they focused their services on post-Katrina activity in particular.

Summary of the Nonprofit Sector Context

Similar to nonprofit organizations across the U.S., Gulf coast nonprofits have experienced declining revenues since the recession, however, Orleans Parish, Louisiana has experienced a much greater decline in revenues comparatively. The numbers of nonprofit organizations have steadily grown across the Gulf region, again, with the exception of Orleans Parish, LA which experienced a growth in numbers during the recession but not to the point of recovering the numbers that existed prior to Katrina. This pattern continues when analyzing the impact on the number of organizations with a focus on service mental health, employment, and community improvement and capacity building needs. Revenues for these kinds of organizations as well as organizations with an environmental focus have increased during the recession, possible a reflection of the continued support for recovery efforts.

When considering the number of nonprofits in each category relative to total revenue, it appears that on average, the organizations are small entities with less than $1 million in annual revenue. This is additional evidence that the capacity of the nonprofit sector in the region is limited.

Appendix B. Summary of Gulf Coast Town Halls

The Deepwater Horizon oil spill was one of the worst manmade environmental disasters ever experienced by the United States. Tens of thousands of people mobilized in the Gulf of Mexico and along the Gulf Coast to combat the spill, plug the well, and mitigate the spill's immediate effects. Early on in the response, President Obama recognized that short-term efforts would not be sufficient to address the spill's enduring effects. In an address to the nation on June 15, 2010, the president appointed Secretary of the Navy Ray Mabus to examine the question of what comes after the well has been sealed and the oil is no longer flowing.

Table 5a. Community Improvement and Capacity Building Focus - Number of Nonprofits

				Percent Change		
	Jul 05	Apr 08	Apr 10	Jul 05 - Apr 08	Apr 08 - Apr 10	Jul 05 - Apr 10
Orleans Parish, LA	303	265	290	-12.5%	9.4%	-4.3%
Gulf Coast	1,728	1,727	1,928	-0.1%	11.6%	11.6%
South	44,921	44,625	47,480	-0.7%	6.4%	5.7%
USA	138,537	136,630	143,159	-1.4%	4.8%	3.3%

Table 5b. Community Improvement and Capacity Building Focus - Total Revenues (in millions)

				Percent Change		
	Jul 05	Apr 08	Apr 10	Jul 05 - Apr 08	Apr 08- Apr 10	Jul 05 - Apr 10
Orleans Parish, LA	$61.07	$68.92	$85.26	12.8%	23.7%	39.6%
Gulf Coast	$235.77	$306.68	$366.69	30.1%	19.6%	55.5%
South	$11,521.9	$15,016.3	$15,330.4	30.3%	2.1%	33.1%
USA	$33,114.1	$41,183.4	$40,035.7	24.4%	-2.8%	20.9%

Table 5c. Community Improvement and Capacity Building Focus - Total Assets (in millions)

				Percent Change		
	Jul 05	Apr 08	Apr 10	Jul 05- Apr 08	Apr 08- Apr 10	Jul 05- Apr 10
Orleans Parish, LA	$120.49	$130.90	$162.25	8.6%	23.9%	34.7%
Gulf Coast	$356.44	$441.96	$505.26	24.0%	14.3%	41.8%
South	$25,7701	$35,785.8	$58,699.7	38.9%	64.0%	127.8%
USA	$79,715.2	$119,443.7	$140,304	49.8%	17.5%	76.0%

Primary Themes

For the past several months, Secretary Mabus has worked to answer that question. In numerous conversations and trips to the Gulf Coast he has met with thousands of residents, including fishermen, health officials, environmental workers, industry leaders, scientists, members of the business community, local leaders, all five state governors, 10 U.S. senators, and every congressman whose district touches the Gulf. During one of his trips, the Secretary convened nine town halls across all five Gulf States during which thousands of residents expressed their opinions on long term recovery.

Though the five states of the Gulf Coast are unique and are affected by Deepwater Horizon in different ways, several common themes and ideas emerged at each of the nine town halls conducted across the Coast:

- The time for action is now. Residents of the Gulf Coast have seen numerous plans for development and environmental recovery fall short of their promised effects. What is needed in the wake of Deepwater Horizon is dedicated restoration funding and rapid deployment of projects geared toward environmental restoration and economic development.
- It is in the Gulf where much of the nation's oil is produced, it is in the Gulf where the damage from the spill occurred, and it is to the Gulf where recovery funds must be dedicated.
- Plans to restore the Gulf Coast must come from the people of the Gulf Coast.
- Recovery plans must be based on good science and research. Colleges and universities across the Gulf Coast have studied the unique ecosystems and economic conditions of the Gulf for years. They have expertise, and are trusted most by Gulf residents, in recommending measures for recovery.
- Gulf residents have high expectations for the restoration effort. However, they also fear that with the spill stopped, the American public will once again forget about the region and the urgent needs it still has. In the wake of devastating hurricanes and two years of acute economic uncertainty, many business owners are concerned they may not make it to next year if help does not arrive soon.
- It was clear to residents that what happens to the ecosystem of the Gulf affects the economy and the welfare of the entire region.
- The Gulf needs to rebuild its image as a place to vacation and as the main source of America's seafood.
- The effect of dispersants on seafood safety, water safety, and beach ecosystems must be openly studied, preferably by Gulf scientists and academic institutions, and the results must be rapidly released to the general public.
- The Gulf economy must be diversified – specifically into clean energy.
- The Gulf needs long-term monitoring to examine enduring effects of the spill on physical and mental health.

What follows is not an exhaustive summary of the town halls, but rather highlights of some of the specific concerns and recommendations expressed by Gulf residents at each of the town halls. These ideas have helped shape Secretary Mabus's report.

Theodore, Alabama: August 2, 2010

In Theodore, the connection between the environment and the economy was of primary concern. As one gentleman said, "Everything that revolves around the coastal areas comes right back to the environment and the health of the environment. But to me, what's first and foremost as to where we go as a region is based on the health of the environment. That's both whether it's tourism, whether it's the fishing industry, you name it. Everything kind of revolves around that."

There was also concern over the lack of information about long-term health impacts associated with a major oil spill. One participant said: "I and my colleagues along the coast have been a bit surprised that when you look at the literature, we really know very little about long-term health effects from an oil well."

Many in the audience expressed a desire for local involvement in restoration planning, as well as an interest in clean energy. Finally, a theme arose that would be a common feature throughout the town hall meetings: the paramount importance of dedicated funds for long-term Gulf restoration. "I've seen so many starts, but then, because there wasn't a dedicated funding source, nothing happened," said one resident. "And so I would urge you to really give that some serious thought."

Robertsdale, Alabama: August 2, 2010

In Robertsdale many people discussed the need for assistance in rebuilding the Gulf Coast seafood and tourism brand. "We are going to need lots of help rebranding the Gulf, if you will," said one participant, "And the seafood is part of our tourism brand, so we need a lot of help in tourism marketing, awareness, education about seafood, all of that type of thing."

Residents also requested support to small businesses, along with an interest in clean energy. "We ought to be able to attract those research and development companies to come here that are going to look at alternative fuel energies," said one participant.

Panama City, Florida: August 3, 2010

In Panama City, the need to restore the Gulf Coast's brand and the importance of small businesses were the primary themes. "Probably our biggest issue right now is not the impact of oil; it's the perception of impact that has really killed tourism for us," said one resident. Another said, "I am looking at these small businesses that are trying to hang on, [and we need to do] anything we can do to help them get from point A to point B."

Participants also mentioned an interest in the clean-energy sector and in the value of community colleges in supporting job retraining.

St. Petersburg, Florida: August 3, 2010

The need for dedicated funds was a major theme in St. Petersburg. "We need a strong funding mechanism to support the many aspects of a Gulf restoration plan," said one resident.

The importance of inclusive restoration planning was also paramount. "We need strong mechanisms and guidelines for an inclusive decision-making process. And not just agencies and not just governors' offices, but we need a clear way for something like a regional citizens' advisory board to provide input and monitor and participate," said another participant.

Additionally, residents expressed concern over the health impacts of dispersants – and over the region's ability to support energy jobs outside the traditional oil-and-gas industries. "We'd like to hear how any kind of long-term restoration plan includes support for other offshore energy projects, not just in the petroleum sector," said one attendee.

Houma, Louisiana: August 5, 2010

In Houma, one woman articulated what the restoration effort must encompass: "Recovery and sustainability for the Gulf Coast depends on three critical resources: the people, the environment and commerce. And we need a Gulf Coast recovery plan that brings the coast back into balance and makes us whole again."

It was made clear that the key to bringing the coast back into balance was the environment, as one woman said: "Our tourism product is our environment and the resources that we pull from that environment, the wonderful fishing, the wonderful birding and the wonderful people, the cultures that live within our environment."

Participants stressed the importance of capitalizing on pre-existing projects for environmental restoration. As one man put it, "Please, let's not reinvent the wheel."

Many attendees mentioned the impact of the spill on Louisiana's seafood brand, and expressed concern over the health impacts of dispersants. Mental health was also a key issue in Houma. Said one participant, "You've heard some of these individuals speak today about the emotional issues that they are dealing with. The fact [is] that some of them have lost their job. They have lost their way of life."

Buras, Louisiana: August 5, 2010

In Buras, the importance of ecosystem restoration was a main concern. Another was the need to coordinate the many disparate programs already in place: "One thing I wanted [to] describe to you is that right now, we have a collection of small programs. It seems to me that's a way to begin to take a lot of the inefficiency out of this. And pulling these programs together, working with existing authorities, and even existing appropriations. Pull these things together."

The importance of local input in recovery planning was on the minds of many, and the damage to the Louisiana seafood brand was a primary concern. Several participants expressed fear that dispersants and oil would keep people from being able to safely fish for years to come.

Galveston, Texas: August 6, 2010

In Galveston, residents emphasized the connection between the Gulf's environment and economy. "We've seen that there's an inextricable link between the health and the environment here in the Gulf and the economy. A healthy Gulf is a vibrant and an economically successful place for the nation," said one.

Participants stated repeatedly that restoring the Gulf's ecosystem is a national issue, and therefore requires dedicated funding from Congress that is not annually appropriated. One man also discussed the importance of the private sector's engagement in restoration: "There really isn't a component for the business community as it relates to economic recovery."

Bay St. Louis, Mississippi: August 7, 2010

In Bay St. Louis, multiple people focused on the need for supporting new business. One woman suggested "incubating new businesses, building new training and innovation centers, and expanding community-college programs."

Job retraining, local hiring for restoration projects, and the opportunity to move people into green jobs was also a top priority: "In terms of short-term economic needs, there are plenty of green-jobs training opportunities out there, getting people with very little training needed to go out and weatherize public buildings, schools."

There was also concern about the impacts of the spill on mental health, and strong interest in citizen engagement in recovery planning.

Ocean Springs, Mississippi: August 7, 2010

In Ocean Springs, a strong case was made for the need to dedicate Clean Water Act Penalties to the Gulf Coast: "I think one of the things that appears at least to be some low-hanging fruit, and it hopefully will be more available and at a quicker level, would be the Clean Water Act violation fines that BP will be responsible for paying."

Another woman built upon this by saying, "We do need the funding to implement a plan. A plan is useless, we've got tons of plans, but we need the funding and the teeth to implement it."

The themes of local involvement; long-term seafood and health monitoring; and transparency were as also raised multiple times, as was interest in clean energy: "Of course we're very interested in renewable energy and less dependence on oil," said one participant.

Appendix C. Tool Kit for Access to Federal Programs

The Economic Assessment teams heard a variety of concerns about access to federal programs, ranging from questions about grant application processes to uncertainty about which federal programs can be utilized. In order to meet these concerns, the federal government has begun developing a recovery "toolkit" which identifies federal programs that may be relevant to communities in different aspects of recovery. The information will be distributed in hard copy across the Gulf, integrated into the disasterrecovery.gov platform, and publicized through a series of technical assistance workshops and webinars in which federal program representatives will provide local governments and other stakeholders an overview of the contents of the tool-kit. Representatives must be available to answer specific questions about different programs and program application processes.

The Contents of the Toolkit will include the following:

I. For Communities and Local Governments
   a. Guide to BP Claims
   b. Guide to NRDA
   c. Recovery Planning Checklist
   d. Guide to Grant Writing
   e. Resource Guide

II. For Fishermen and Consumers
   a. Federal Fisheries Re-opening Decisions and Other Information
   b. State Fisheries Re-opening Information and Seafood Safety
   c. Ensuring the Safety of Your Seafood
   d. Fishery Closure Frequently Asked Questions

III. For Individuals
   a. Guide to GCCF Claims
   b. Guide to Private Lands Remediation
   c. Shoreline and Habitat Fact Sheet

IV. For NGOs
   a. Guide to GCCF Claims

End Notes

1 See http://www.ustravel.org/sites/default/files/page/2009/11/GulfOilSpillAnalysisOxford Econo mics710.pdf

2 See http://www.eia.doe.gov/special/gulfofmexico/index.cfm

3 Bureau of Ocean Energy Management, Regulation and Enforcement Chart, "Total Federal Offshore Reported Royalty Revenues," http://www.mrm.boemre.gov/MRMWebStats/Fed OffReportedRoyaltyRevenues.aspx?yeartype=FY&year=2009&d ateType=AY.

4 State of Louisiana Press Release, "Governor Jindal Announces Funding for Fishery Resource Monitoring Program, Calls on BP to Fund Long-Term Seafood Plan, August 19, 2010, www.emergency.louisiana.gov

5 This report is not meant to suggest that any particular funds are available for any particular project but again, is meant to identify the needs in the Gulf, provide some possible restoration solutions and to propose a structure that will coordinate and enhance ongoing efforts.

6 U.S. Scientific Teams Refine Estimates of Oil Flow from BP's Well Prior to Capping." Restore the Gulf.Gov: Deepwater Horizon Incident Joint Information Center. 2 August 2010. http://app.restorethegulf.gov/go/doc/2931/840475/.

7 Ibid.

8 NOAA Administrator Jane Lubchenco, transcript of August 27, 2010 teleconference ("NOAA Administrator Briefing to Discuss the Reopening of Federal Waters Along Western Louisiana," www.restorethegulf.gov)

9 The daily official "Operations and Ongoing Response" briefing document at www.restorethe gulf.gov

10 "At the height of the Deepwater Horizon Response, nearly four million feet of boom, response-wide, was deployed. ..." Source: Press Release, Deepwater Horizon Incident Joint Information Center, "More Than 1.5 Million Feet of Hard Boom Recovered from Coastal Waters," August 23, 2010. http://www.deepwaterhorizon response.com/go/doc/2931/872245/

11 "BP Deepwater Horizon Oil Budget: What Happened to the Oil?", Jane Lubchenco, Marcia McNutt et al., August 4, 2010, http://www.noaanews.noaa.gov/stories2010/20100804 oil.html.

12 Unless otherwise indicated, the data and conclusions cited throughout this report from state and other non- federal research studies or sources have not been separately verified or adopted by the federal government.

13 "Hypoxia in the Gulf of Mexico: Progress towards the completion of an Integrated Assessment," National Oceanic and Atmospheric Administration, Retrieved September 21, 2010, http://oceanservice.noaa.gov/products/pubs_hypox.html

14 ""NOAA's Gulf of Mexico Region," Retrieved September 23, 2010, http://www.ppi.noaa.gov/Regional_Collaboration/Regional_Overviews/GulfofMexicoRegionOverview_042407.pdf

[15] John Tibbetts, Louisiana's Wetlands: A Lesson in Nature Appreciation," Environ Health Perspect., January 2006; 114(1): A40–A43, National Institute of Environmental Health Sciences, Retrieved September 21, 2010, http://www.ncbi.nlm.nih.gov/pmc/articles/ PMC1332684/

[16] "Species Listed in the Fishery Management Plans of the Gulf of Mexico Fishery Management Council," Gulf of Mexico Fishery Management Council, Retrieved September 21, 2010, http://www.gulfcouncil.org/Beta/ GMFMCWeb/downloads/species%20managed.pdf

[17] "Hypoxia," Mississippi River Gulf of Mexico Watershed Nutrient Task Force, Retrieved, September 21, 2010, http://water.epa.gov/type/watersheds/named/msbasin/gulfnews.cfm

[18] "U.S. Scientific Teams Refine Estimates of Oil Flow from BP's Well Prior to Capping," RestoretheGulf.gov, Deepwater Horizon Incident Joint Information Center, Retrieved, September 21, 2010, http://app.restorethegulf.gov/go/doc/2931/840475/

[19] "Deepwater Horizon/BP Oil Spill: Size and Percent Coverage of Fishing Area Closures Due to BP Oil Spill," http://sero.nmfs.noaa.gov/ClosureSizeandPercentCoverage.htm

[20] "Mobile Bay National Estuary Program," Retrieved, September 21, 2010, http://www.mobilebaynep.com/site/ estuary/description.htm.

[21] Alabama Department of Conservation and Natural Resources,"The Forever Wild Land Trust – An Interim Report to the Citizens of Alabama, 1992-2009."

[22] News Release, September 3, 2009, Texas Parks and Wildlife Department, Retrieved September 21, 2010, http://www.tpwd.state.tx.us/newsmedia/releases/?req=20090903a

[23] National Marine Fisheries Service, 2010, "Fisheries Economics of the United States," 2008. U.S. Dept. Commerce, NOAA Tech. Memo. NMFS-F/SPO-109, Retrieved September 21, 2010, See http://www.st.nmfs.noaa.gov/st5/publication/index.html

[24] U.S. Army Corps of Engineers, 2010, "Coastal Wetlands Planning, Protection, and Restoration Act," Retrieved September 21, 2010, http://www.mvn.usace.army.mil/pd/cwppra_mission. htm.

[25] "Coastal Wetlands Planning, Protection, and Restoration Act," USGS National Wetlands Research Center, Louisiana Coastal Wetlands Planning Protection and Restoration Act Program Website, Retrieved September 21, 2010, http://www.lacoast.gov/new/About/ Default.aspx

[26] When the term "health" is used, it is defined as a state of complete physical, mental and social well-being, and thus includes behavioral health in all instances.

[27] Public health surveillance is the ongoing and systematic collection, analysis, interpretation, and dissemination of data about a health-related event for use in public health actions to reduce morbidity and mortality, improve health, and provide necessary financial and social support to individuals and families in need.

[28] Resources to track Health Professional Shortage Areas (HPSAs) and Medically Underserved Areas/Populations (MUA/P) can be found at the HHS Health Resources and Services Administration website: http://www.hrsa.gov/data-statistics/index.html.

[29] See http://www.iom.edu/Reports/2010/Assessing-the-Effects-of-the-Gulf-of-Mexico-Oil-Spill-on-HumanHealth. aspx.

[30] Resources to track Health Professional Shortage Areas (HPSAs) and Medically Underserved Areas/Populations (MUA/P) can be found at the HHS Health Resources and Services Administration website: http://www.hrsa.gov/data-statistics/index.html.

[31] See http://www.iom.edu/Reports/2010/Assessing-the-Effects-of-the-Gulf-of-Mexico-Oil-Spill-on-HumanHealth. aspx.

[32] Dyer, CL, Gill, DA and Picou, S. (1992). *Social disruption and the Valdez oil spill: Alaskan natives in a natural resource community.* Sociological Spectrum, 12, 105-126.

[33] McNally, TS (1997) *Technological disaster and chronic psychological stress: An evaluation of the conservation of resources stress model.* Sociological Spectrum, 17, 87-92.

[34] See http://www.iom.edu/Reports/2010/Assessing-the-Effects-of-the-Gulf-of-Mexico-Oil-Spill-on-HumanHealth. aspx.

[35] Fothergill, A., Maestas, E., and Darlington, J. D. (1999), Race, ethnicity and disasters in the United States: A review of the literature, *Disasters 23*(2), 156-173; Miller, K. S. and Simile, C. (1992), *"They Could See Stars From Their Beds": The plight of the rural poor in the aftermath of Hurricane Hugo.* Boulder, CO: Westview Press; Peacock, W.G., Morrow, B.H., and Gladwin, M. (1997), *Hurricane Andrew: Ethnicity, gender, and the sociology of disasters.* New York: Routledge; Phillips, B.D. (1993), Cultural diversity in disasters: Sheltering, housing, and long term recovery. *International Journal of Mass Emergencies and Disasters 11* (1): 99-110; Rovai, E. (1994), The Social geography of disaster response: Differential community response to the North Coast Earthquakes. *Association of Pacific Coast Geographers Yearbook* 56; The White House (2006), *Federal Response to Hurricane Katrina: Lessons Learned* (p.59).

[36] See http://www.iom.edu/Reports/2010/Assessing-the-Effects-of-the-Gulf-of-Mexico-Oil-Spill-on-HumanHealth. aspx.

[37] Before, during, and after an incident, members of at-risk populations may have additional needs in one or more of the following functional areas: communication, medical care, maintaining independence, supervision, and transportation. In addition to children, senior citizens, and pregnant women, individuals who may need

additional response assistance include those who have disabilities, live in institutionalized settings, are from diverse cultures, have limited English proficiency or are non-English speaking, are transportation disadvantaged, have chronic medical disorders, and have pharmacological dependency.

[38] See http://www.samhsa.gov/Disaster/traumaticevents.aspx.

[39] Monthly reports April-June 2010 from the National Domestic Violence Hotline to the Family Violence Prevention and Services Program, ACF.

[40] See http://www.cdc.gov/niosh/topics/oilspillresponse/gulfspillhhe.html.

[41] Health security, as envisioned in the National Health Security Strategy, is a state where people and communities are prepared for, protected from, and resilient in the face of health threats or incidents with negative health consequences.

[42] The Behavioral Risk Factor Surveillance System (BRFSS) is a state-based system of health surveys that collects information on health risk behaviors, preventive health practices, and health care access primarily related to chronic disease and injury. More than 350,000 adults are interviewed each year, making the BRFSS the largest telephone health survey in the world. States use BRFSS data to identify emerging health problems, establish and track health objectives, and develop and evaluate public health policies and programs.

[43] Picu, SJ, and Arata, CM (1999) Chronic psychological impacts of the Exxon Valdez oil spill: resource Loss and Commercial Fishers. Sociological Spectrum, 23, 12-19.

[44] Institute of Medicine, *Preventing Mental, Emotional, and Behavioral Disorders Among Young People: Progress and Possibilities*, 2009.

[45] Norris, FH, Friedman, MJ, Watson, PJ, Bryne< CM, Diaz, E., and Kaniasty, K (2002). 6000 Disaster victims speak: Part 1. An empirical review of the empirical literature, 1981-2001. Psychiatry, 65(3), 2007-226.

[46] Watson, P., Thim, K., Santiago, P. CCP Model Analysis: Informing the Program through Evidence and Expert Consensus; An Interdisciplinary Task Force White Paper. Publication pending.

[47] Uninsured rates in the region are as follows: Texas (ranks #1 in the country) at 25.2% of the state's population; Florida (#3) at 20.2%; Louisiana (#5) at 19.3%; Mississippi (#8) at 18.4%; Alabama (32) at 12%.

[48] Chronic Disease Prevalence (Diabetes, Cardiovascular Disease, Disability) and Health Care Coverage-Behavioral Risk Factor Surveillance System 2009; Resources to track Health Professional Shortage Areas (HPSAs) and Medically Underserved Areas/Populations (MUA/P) can be found at the HHS Health Resources and Services Administration website: http://www.hrsa.gov/data-statistics/index.html

[49] EPA Technical Product Bulletins #D-4 (COREXIT EC9500A), D#1 (COREXIT EC9527A); ATSDR, Toxicological Profile for Total Petroleum Hydrocarbons (TPH) DHHS, Editor. 1999: Atlanta GA; CDC-emergency.cdc.gov/gulfoilspill2010/light_crude_health_ professionals. asp.

[50] See http://www.iom.edu/Reports/2010/Assessing-the-Effects-of-the-Gulf-of-Mexico-Oil-Spill-on-HumanHealth. aspx.

[51] See http://www.iom.edu/Reports/2010/Assessing-the-Effects-of-the-Gulf-of-Mexico-Oil-Spill-on-HumanHealth. aspx.

[52] Rodriguez-Trigo G, et al. *Health Changes in Fishermen 2 Years After Clean-up of the* Prestige *Oil Spill,* Ann Int Med. 2010; 153(6); Solomon GM Janssen S *Health Effects of the Gulf Oil Spill.* JAMA. 2010; 304(10):1118-1119.

[53] Solomon GM Janssen S *Health Effects of the Gulf Oil Spill.* JAMA. 2010; 304(10):1118-1119.

[54] Information on this grant opportunity can be found at the following website: https://grants.hrsa.gov/ webExternal/FundingOppDetails.asp?FundingCycleId=450970E7-563E-4D2D-A021-5C775F7F614E&View Mode=EU&GoBack=&PrintMode=&Online AvailabilityFlag=&pageNumber=&version=&NC=&P opup=.

[55] See http://www.mobilehealthclinicsnetwork.org/partners.html for a listing of a Healthcare Partners Collaborative to the Mobile Health Clinics Network organization and the related *2010 Sixth Annual Forum- Mobile Healthcare: At the Crossroads of Opportunity* scheduled in New Orleans September 25 – 28.

[56] Fothergill, A., Maestas, E., and Darlington, J. D. (1999), Race, ethnicity and disasters in the United States: A review of the literature, *Disasters 23*(2), 156-173; Miller, K. S. and Simile, C. (1992), *"They Could See Stars From Their Beds": The plight of the rural poor in the aftermath of Hurricane Hugo.* Boulder, CO: Westview Press; Peacock, W.G., Morrow, B.H., and Gladwin, M. (1997), Hurricane *Andrew: Ethnicity, gender, and the sociology of disasters.* New York: Routledge; Phillips, B.D. (1993), Cultural diversity in disasters: Sheltering, housing, and long term recovery. *International Journal of Mass Emergencies and Disasters 11* (1): 99-110; Rovai, E. (1994), The Social geography of disaster response: Differential community response to the North Coast Earthquakes. *Association of Pacific Coast Geographers Yearbook* 56; The White House (2006), *Federal Response to Hurricane Katrina: Lessons Learned* (p.59).

[57] Report from Louisiana State Domestic Violence Coalition to the Family Violence Prevention and Services Program, ACF

[58] Any expenditures under such programs must be consistent with the program's underlying statute and any relevant administrative documents such as funding and award announcements.

[59] See http://www.acf.hhs.gov/programs/ofa/policy/pi-ofa/2005/pi2005-6.htm.

[60] See http://www.acf.hhs.gov/programs/ofa/recovery/tanf-faq.htm.

[61] http://www.hhs.gov/gulfoilspill/bpfundinglevels.html.

[62] http://www.fns.usda.gov/snap/ebt/default.htm.

[63] http://www.iom.edu/Reports/2010/Assessing-the-Effects-of-the-Gulf-of-Mexico-Oil-Spill-on-HumanHealth.aspx.

[64] Aguilera et al., Review of the health effects of exposure to spilled oils on human health, Appl Toxicol. 2010, 30:291-301.

[65] Collection and analysis of biospecimens, such as urine and blood, could be useful for deriving biomarkers of exposure and early adverse response, identifying highly susceptible individuals and subgroups, and informing treatment, prevention, and future surveillance activities following oil spills. A pool of banked specimens could be established for future analysis. Parallel lab-based toxicological studies in animals and cell systems should be conducted to identify potential hazards posed by oil and dispersant chemicals from the Gulf that could be addressed in surveillance or long-term research studies.

[66] GDP data from Bureau of Economic Analysis, http://www.bea.gov/regional/gsp/. Employment data based on Quarterly Census of Employment and Wages, http://www.bls.gov/cew/. Much data such as state GDP data, travel and tourism spending and employment data, and commercial fishing landings data were only available through 2008. For consistency, 2008 data are also used for other statistics appearing in this section.

[67] NOAA-defined coastal counties are the coastal counties that NOAA recommends for the Bureau of the Census' *StatisticalAbstract* publication series. There are 139 Gulf of Mexico coastal counties (not including GA). See http://www.census.gov/geo/landview/lv6help/ coastalcty.pdf.

[68] Tourism employment is difficult to measure because many of the businesses that sell goods and services to visitors do the same to non-visitors. The Bureau of Economic Analysis attempts to isolate employment directly related to visitors with their Travel and Tourism Satellite Account data, which is the data we rely on here.

[69] This includes more counties than just those adjacent to the Gulf of Mexico. The employment figure would drop to about 301,000 focusing on only those counties directly adjacent to the Gulf of Mexico.

[70] State specific totals are LA (107,000), MS (20,000), AL (20,000), FL (209,000) and TX (232,000).

[71] The average tourism worker in these coastal counties earned about $30,000 in 2008.

[72] For more detail on these estimates, please see the Appendix.[72] This total does not include people employed within the retail and wholesale seafood sectors of the economy.

[73] Average annual wages represent weighted averages across the five Gulf Coast states. State annual wage data obtained from Quarterly Census of Employment and Wages. Weighting is based on state employment within each industry. QCEW does not report an annual wage for harvesting for the state of Mississippi, so the average across Louisiana and Alabama was used instead.

[74] Quarterly Census of Employment and Wages, including NAICS code 213111 Drilling Oil and Gas Wells and NAICS code 213112 Support Activities for Oil and Gas Operations.

[75] The data does not allow distinction between offshore and onshore oil and gas drilling. A recent report released by the Administration conducted a detailed analysis of the impact of the moratorium imposed after the Spill. For more information please see: http://www .esa.doc.gov/drillingmoratorium.pdf

[76] Throughout the initial claims process, the Administration consistently advocated for greater transparency into claims data. On June 8, 2010, for example, National Incident Commander Thad Allen wrote a letter to BP CEO Tony Hayward stating, "we need additional information [on claims processing] to assess how well the process is meeting the critical needs of individuals, families, and businesses whose livelihoods are being impacted by the spill." http://www.deepwaterhorizonresponse.com/posted/2931/NIC Letter to BP CEO.62 1247. pdf.

[77] Advanced Resources International, Inc. *Economic Impact of Oil Spill: Spill Unit Costs for Tankers, Pipelines, Refineries, and Offshore Facilities*. Prepared for the United States Department of Energy, 1993. See also, "An Assessment of the Impact of the Exxon Valdez Oil Spill on the Alaska Tourism Industry," McDowell Group, August 1990.

[78] Cedre: Amoco Cadiz, A Year Later, First Consequences. http://www.cedre.fr/en/spill/amoco/ amoco1year.php (accessed August 20, 2010). See also, National Oceanic and Atmospheric Administration, "Assessing the Social Costs of Oil Spills: The *Amoco Cadiz* Case Study," July 1983.

[79] Grigalunas, Thomas A., et al., "Estimating the Cost of Oil Spills: Lessons from the Amoco Cadiz Incident," *Marine Resource Economics*, Vol. 2, No. 3, 1986: 239-262 and National Oceanic and Atmospheric Administration, "Assessing the Social Costs of Oil Spills: The *Amoco Cadiz* Case Study," July 1983.

[80] Mead, W.J. and P.E. Sorensen, "The Economic Cost of the Santa Barbara Oil Spill," *Santa Barbara Oil Spill Symposium: An Environmental Inquiry*. United States Geological Survey, 1970.

[81] Garcia Negro, M.C., et al., "Estimating the Economic Impact of the Prestige Oil Spill on the Death Coast (NW Spain) Fisheries," *Marine Policy*, Vol. 33, 2009: 8-23.

[82] Cohen, M.J., "Technological Disasters and Natural Resource Damage Assessment: An Evaluation of the Exxon *Valdez* Oil Spill.

[83] Exxon *Valdez* Oil Spill Trustee Council, Commercial Fishing. http://www.evostc.state.ak.us/ recovery/ lingeringoil.cfm (accessed August 20, 2010).

[84] See http://www.acf.hhs.gov/programs/ofa/tanf/apprTANFemerfund.html.

[85] See http://www.whitehouse.gov/blog/2010/08/17/what-you-missed-open-questions-gulf- sea food-safety-withdr-lubchenco.

[86] See http://www.restorethegulf.gov/travel.shtm.

[87] Recent sources describe between $87 million and $89.5 million in tourism grants for the affected states. Data provided to the House Energy and Commerce Committee cites the latter figure (http://energycommerce. house.gov/documents/20100901/Castor.BP Advertising.2010.9.1.pdf), while a recent public report from BP reflects $87 billion to the four states.

[88] See http://www.bp.com/genericarticle.do?categoryId=2012968&contentId=7062187.

[89] See http://www.dep.state.fl.us/deepwaterhorizon/files/091510 back to beach.pdf.

[90] See http://www.floridagulfresponse.com/go/doc/3059/897671/.

[91] See http://www.floridagulfresponse.com/go/doc/3059/897671/.

[92] In Florida, a September 30 deadline was established in a Memorandum of Understanding between the participating state tourism entities and BP: http://www.myflorida house. com/FileStores/Web/HouseContent/ Approved/Announcements/uploads/documents/ responseworkgrou p/fi na l/Workgrou p%202%20-%20Fi na l%20Report%20(with% 20attach ments). pdf, 2.

[93] For example, Escambia County, FL reported to its Assessment and Evaluation team that it was having difficulty spending funds that it had received, while neighboring Santa Rosa reported the opposite experience, as described above.

[94] AL 9; FL 20; LA 12; MS 15.

[95] This assistance was in the context of immediate disaster relief and will not continue throughout the claims process. Moving forward, the recommendations in this section will be critical to helping small businesses navigate the claims process.

[96] See Federal Procurement Data System Gulf Oil Spill Report available at https://www. fpds.gov/fpdsng cms/.

[97] SCORE stands for Service Corps Of Retired Executives.

[98] According to the 2009 State Energy Efficiency Scorecard from the American Council for an Energy Efficient Economy.

[99] See http://www.aceee.org/ sector/state-policy/scorecard.

[100] See http://www1.eere.energy.gov/biomass/pdfs/algalbiofuelsroadmap.pdf.

[101] See http://www.eia.gov/dnav/pet/petpnpcap1a(na)8D0BpCDa.htm.

[102] See http://www.dailyfinance.com/story/federal-grant-algae-fuel-research/19534858/.

[103] See http://www1.eere.energy.gov/biomass/news detail.html?news id=16311.

See http://www.energy.gov/news2009/8519.htm.

[104] See http://www.epa.gov/diesel/newsroom.htm.

[105] See http://www.aapa-ports.org/Industry/contentcfm?ItemNumber=900&navItemNumber=551.

[106] In addition Penascal Wind Power, LLC, in Sarita, TX received $114.1 million and Inadale Wind Farm, LLC, in Roscoe received $94.2 million for wind projects. See http://energy. gov/recovery/documents/Recovery Act Memo Texas. pdf.

[107] See http://energy.gov/recovery/documents/RecoveryActMemoLouisiana.pdf.

[108] See http://energy.gov/recovery/documents/RecoveryActMemoFlorida.pdf.

[109] See http://energy.gov/recovery/documents/RecoveryActMemoFlorida.pdf.

[110] See http://energy.gov/recovery/documents/RecoveryActMemoLouisiana.pdf.

[111] Nonprofit organizations are defined as tax exempt, charitable organizations and include voluntary, faith-based and community organizations, charities, foundations and philanthropic groups as well as professional associations and academic institutions.

[112] In order to determine the needs and planning for a coordinated response to community issues emerging from the oil spill, the Corporation for National and Community Service (CNCS) analyzed information to evaluate the state of nonprofits on the Gulf Coast. The resulting conclusions provide valuable insight into the nonprofit sector's potential capacity and likely limitations in responding to the oil spill and recovery effort, as well as recommendations for actions and initiatives that will allow the nonprofit sector to fill gaps not met by government services.

[113] Wing, Kennard T., Katie L. Roeger, and Thomas H. Pollak. 2009. "The Nonprofit Sector in Brief: Public Charities, Giving, and Volunteering, 2009." The Urban Institute: Washington, DC. Available at http://www.urban.org/UploadedPDF/412085-nonprofit-sector-brief.pdf. As seen in Figure 2, page 3, government accounts for 29.8 percent of revenues across all reporting charities.

[114] Based on March 2010 survey of 1,300 nonprofit leaders conducted by the Nonprofit Finance Fund. Summary of results accessed at: http://nonprofitfinancefund.org/ announcements/2010/nonprofit-finance-fund-survey-americas-nonprofits-brace-tough-2010

[115] The Urban Institute's National Center for Charitable Statistics reports data from nonprofit tax returns in their IRS Business Master Files (BMFs). The BMFs from July 2005, April 2008, and April 2010 are the source for this statistic, as well as the detailed statistics reported in Appendix B.

[116] According to the April 2010 BMF, 55.2 percent of all Gulf Coast reporting nonprofits have revenues less than $1 million.

[117] Each of the states conducted a survey asking nonprofit organizations to indicate if they had seen an increase in need for services or the provision of services since the oil spill.

[118] http://gulfcoastfund.org/bp-oil-disaster.html

[119] Since the most recent BMF contains data that were submitted on or before April 2010, it allows us to characterize the nonprofit sector in the Gulf Coast region during the year-long period immediately before the oil spill. The April 2010 data on Gulf Coast nonprofits also reflects the lingering effects of two earlier crises: Hurricane Katrina, which struck in August 2005, and the national economic downturn of the last few years.

[120] The coastal areas include counties that are declared by the Small Business Administration to be economic injury areas. The Gulf Coast regional statistics are based on data reported by nonprofits headquartered in the following counties: Baldwin and Mobile counties in Alabama; Bay, Collier, Escambia, Gulf, Lee, Monroe, Santa Rosa, Okaloosa, Wakulla, and Walton counties in Florida; Cameron, Iberia, Jefferson, LaFourche, Orleans, Plaquemines, Saint Bernard, Saint Mary, Saint Tammany, Terrebonne, and Vermilion parishes in Louisiana; and George, Hancock, Harrison, Jackson, Pearl River, and Stone counties in Mississippi.

[121] The South Region, as defined by the Census Bureau, consists of Alabama, Arkansas, Delaware, the District of Columbia, Florida, Georgia, Kentucky, Louisiana, Maryland, Mississippi, North Carolina, Oklahoma, South Carolina, Tennessee, Texas, Virginia, and West Virginia.

[122] These results, while suggestive, do not allow us to isolate the impacts of Katrina,and the national economic downturn on revenues for Gulf Coast-area or even Orleans Parish nonprofits.

[123] The financial data from the BMFs do not contain information about the types of requests for assistance that nonprofits receive, nor how much assistance they provide. The BMF data does, however, categorize nonprofits based on their main activities, based on the NTEE For details on the NTEE (National Taxonomy of Exempt Entities) codes, please see http://nccs.urban.org/classification/index.cfm.

[124] For this section, we focused on organizations who identified as being one of the following types of organizations using the NTEE (National Taxonomy of Exempt Entities) codes: Animal Related; Mental Health; Crime, Legal Related; Employment, Job Related; Food, Agriculture, Nutrition; Housing, Shelter; Public Safety, Disaster Preparedness; Youth Development; Human Services, Multipurpose and Other; Civil Rights/Advocacy; Community Improvement; Philanthropy, Voluntarism; and Science and Technology. For details on the NTEE codes, please see http://nccs.urban.org/classification/index.cfm.

[125] Based on 2010 data, there are 19,000 Mental Health and Crisis Intervention organizations nationally.

[126] Based on 2010 data, there are 36,000 environmental organizations nationally.

[127] Almost 40,000 nonprofit organizations (out of almost 1.6 million) are categorized as having a main focus of employment, as of April 2010.

INDEX

C

D

E

F

G

H

I

J

K

L

M

N

Q

R

S

V

W

Y